ORIGINS OF HUMAN SOCIALIZATION

ORIGINS OF HUMAN SOCIALIZATION

DONALD W. PFAFF
Laboratory of Neurobiology and Behavior, Rockefeller University, New York, NY, United States

ELSEVIER

ACADEMIC PRESS
An imprint of Elsevier

Academic Press is an imprint of Elsevier
125 London Wall, London EC2Y 5AS, United Kingdom
525 B Street, Suite 1650, San Diego, CA 92101, United States
50 Hampshire Street, 5th Floor, Cambridge, MA 02139, United States
The Boulevard, Langford Lane, Kidlington, Oxford OX5 1GB, United Kingdom

Library of Congress Cataloging-in-Publication Data
A catalog record for this book is available from the Library of Congress

British Library Cataloguing-in-Publication Data
A catalogue record for this book is available from the British Library

ISBN 978-0-323-85861-8

Publisher: Nikki Levy
Acquisitions Editor: Joslyn Chaiprasert-Paguio
Editorial Project Manager: Kristi Anderson
Production Project Manager: Omer Mukthar
Cover Designer: Julia Heffernan

Typeset by SPi Global, India

Contents

About the author

I knew that I was interested in brain and human behavior, starting in tenth grade English class when we were studying a novel involving character formation. But in my high school area in a small city, if you were really good in science and math and you were a boy, you were "supposed" to go into math and chemistry and physics—the "hard" sciences.

At Harvard I started in physics and math until, after my father (a surgeon) died, I really had to determine what truly would interest me as my life's work. An inspiring teacher Paul Rozin propelled me toward neurobiology and behavior. As a volunteer at a Massachusetts state mental hospital, on the back wards I witnessed more than one type of schizophrenia. Most of all, I saw that the standard of medical care by the psychiatrists in that mental hospital would not live up to my father's medical standards. Therefore, I would go for a PhD in research, not an MD.

In graduate school at MIT I walked into the lab of Joseph Altman, born in Hungary and succeeding in America, and knew that I had to look no further. His use of histochemical techniques achieved a level of precision and beauty I had never imagined. He had discovered that the dogma "all of our neurons are born by the time of birth" was simply wrong. But instead of just helping him follow that up, I said (referring to my experience as a psychiatric helper) "Joe, I would like to study something closely related to behavioral function". He gave me a seven word answer "I'd like to do something with hormones". His answer led to my working out the first set of mechanisms for a vertebrate animal's behavior, a simple sex behavior.

The choice of problem was crucial. Having discovered hormone receptors in the brain, I was able to aim toward solving a behavior, female lordosis behavior that (a) was a simple enough problem to solve, (b) was the behavior that controlled reproduction, and (c) united mammalian behavioral biology with molecular endocrinology.

Accomplishing a lot in the lab with sexual and maternal behaviors led me to think, in my spare time, about broader issues of human behavior.

My *Man and Woman: An Inside Story* separated fact from fiction about sex differences in behavior. And *The Altruistic Brain* claims that altruistic behaviors can be explained on the basis of known brain mechanisms—no philosophy required.

Always interested in the implications of science in my field for the human condition—I had narrowly decided not to go to medical school—I had cross-registered from MIT as a graduate student into all the neurobiology and endocrinology courses at Harvard Medical School. There I could appreciate the precision and beauty of experimental work, now in the electrophysiological realm, of Steve Kuffler. But also I could see how the brilliant professors there could translate their basic work into human biology and medicine.

That background gave me courage, the nerve, to attempt *Origins of Human Socialization*. This book deals with subjects a thousand times more sophisticated than my laboratory work on simple sex and maternal behaviors. The book wants to answer the question "where does the human spark for social contact come from." Knowing that archeologists and anthropologists had made great strides toward answers, and that developmental psychologists could chart babies' social engagement from the get-go, I was loathe to limit my writing to my specialty, neuroscience.

So the concept I developed to write this book is that the expression of the most basic human instinct toward social contact depends on three roots: (i) Evolution—but what is it that evolves? (ii) What evolves are the Developmental Psychological capacities to seek social contacts of increasing complexity. (iii) And governing those social contacts in younger and older animals and people are the Brain Mechanisms which my colleagues work on. These three areas of study do not compete with each other. Instead the three roots are intertwined in the manner that they serve to support the human instinct toward social contact.

Together, these approaches create a multi-disciplined, unified concept of how our drive toward prosocial activity evolved and came to govern our behavior. I do not try to account for all social behavior, but for the initial, instinctive "spark" that makes all other social behavior possible. I ask why we ineluctably move toward other human beings and why, in some instances, the instinct fails.

I tell the story from a personal point of view—I know and am friendly with some of the authors quoted. But for a working scientist the story rarely stops just here. Most chapters introduce new, unanswered questions, and all of the chapters have lots of Further Reading.

Acknowledgments

The logic, interest and readability of this book all are due to the criticism and assistance of Sandra Sherman, J.D., PhD. Alexandra Kramskaya and Julia Heffernan created the figures.

The chapter on fish was written with the help of expert neuroendocrinologist Ishwar Parhar, director of the Brain Research Institute at the Monash University campus in Kuala Lumpur, Malaysia. As well, Chapter 9, on autism, is adapted from the Trends in Neuroscience article written with Boston University School of Medicine professor Helen Barbas.

Material on sex difference in autism was researched by psychiatrist Hagar Kandel, M.D. in my lab. And University of Miami professor Daniel Messinger was kind enough to critique the writing on autism.

Part of the chapter on Elementary Social Responses was adapted from Pfaff, Tabansky and Haubensak (2019). In turn, that thinking benefitted from comments of professors Dayu Lin (NYU) and Gabriela Gonzalez-Mariscal (CINVESTAV, Mexico).

Several sources of knowledge and expertise added to the quality of the book. Professors Rui Oliveira (Lisbon, ISPAT) and Gil Levkowitz (Weizmann Institute, Israel) offered advice with respect to the chapter on fish. Further, authors in my review volume edited with professor Marion Joels (University of Utrecht Medical Center, the Netherlands) provided useful sources of relevant information. Conversations with professor emeritus James Darnell (Rockefeller) led to my thinking and writing in Chapter 11.

It goes without saying that I am grateful to all these scholars.

Introduction

> *Man is a sociable being and it is for aught I know one of the worst of punishments to be excluded from society. I have read an abundance of fine things on the subject of solitude and I know 'tis a common boast in the mouths of those that affect to be thought wise that they are never less alone than when alone.' I acknowledge solitude an agreeable refreshment to a busy mind; but were these thinking people obliged to always be alone, I am apt to think they would quickly find their very being unsupportable to them.*

Benjamin Franklin wrote these much-quoted lines in his Journal when he was 21, crossing from London to Philadelphia, observing life from the close quarters of an 18th-century ship. Apparently, a fellow passenger was caught cheating at cards and, when he refused to pay the fine, he was shunned. But on a long voyage the isolation was too much even for this anti-social card-sharp and he finally paid up, leaving Franklin to affirm that the man found his own company "unsupportable" without the company of others.

As an Enlightenment figure, Franklin could have predicted this outcome. In his great biography of Franklin, Walter Isaacson notes that it was a fundamental Enlightenment tenet that there was a "sociable affinity, based on the natural instinct of benevolence, among fellow humans" which Franklin exemplified. My argument is that however much Enlightenment thinkers believed in a natural affinity among human beings, we can now prove it and demonstrate where it comes from. Like Franklin and his brethren, we continue to look at nature for the source of this affinity, but we do so scientifically. Where they were right based on keen observation of the human condition, we are now confident based on all the tools of an advanced, technically sophisticated society.

It is crucial to understand the compelling evidence for our natural sociable affinities since with some of the gruesome anti-sociability of the 20th and 21st centuries—magnified by coronavirus rules—it is easy to conclude that our natural affinities point in the opposite direction. Take, for example,

Orson Welles' famous statement: "We're born alone, we live alone, we die alone. Only through our love and friendships can we create the illusion for the moment that we're not alone." Welles is suggesting that even the semblance of a human connection is illusory, much like the illusions that he created for a living and then made to disappear (consider "War of the Worlds"). Shunning even the illusion of connection, Otto Weininger claimed: "Man is alone in the world, in tremendous eternal isolation. He has no object outside himself..." The vast pessimism of this statement, potentially appealing on some philosophical level where the point is to explain a recurrent weltschmerz, is still the product of philosophy. But we have to assess the human condition as it actually is, i.e., beyond illusion, speculation, and philosophy—that is, as science explains it. This book seeks to do that, and to demonstrate why, based on the latest science, human beings (like many species) have evolved to display an intense affinity for each other.

By "affinity" I mean a spontaneous natural liking or sympathy for another person, an interest in forming a bond or relationship. Of course, these are broad categories, and human affinities take on all kinds of shapes—friendships are affinities, but so is an impulse to save another human *because* they are human. We are drawn to our own kind. More than just wanting company, we have an irresistible need to populate our mental and physical space with other people. Sometimes, these are special people: parents, for example, have a strong affinity for their children, and vice versa. The attraction that a man feels for a woman is an affinity. We read *People Magazine*, or the Style section of the *New York Times*, because curiosity expresses an affinity. We just have to know! Indeed, I could go on naming more affinities because affinities are everywhere. That is my whole point. Here in *Origins*, I am explaining the commonplace. What's different, however, is that while the phenomenon is obvious and ubiquitous, I think the explanation for why it exists is anything but. There are complex neurobiological, evolutionary, and developmental reasons for why we are drawn to one another, and that is what I plan to discuss here.

I will argue that we connect to each other instinctively, and that all that mutual interest can be thoroughly demystified. Maybe that's disappointing. But the "spark" that I cite in the title is not something ineffable. We may like to think of it that way (as in "Some Enchanted Evening"), but it's readily understandable based on the data. Of course, that's not to say that the explanation is unexciting. I pursued it for years because, with all the galloping developments in science, I found (and find) it immensely exciting.

Three launching platforms

My argument is that the origin of our instinct to "connect" is evident in three independent empirically-based knowledge bases:

- First, paleoanthropology—the shift from hunter-gathers to city dwelling sheds light on the development of social habits.

- Second, the study of newborns—infants develop quickly from an inability to focus on faces to recognizing their mothers; brain mechanisms that favor social habits grow rapidly.
- Third, neuroscience—the effect of genes, hormones, and brain structure on human connectivity.

Some guardians of their own disciplines might regard these three approaches to human social behavior as comprising a zero-some game: paleoanthropology versus developmental psychology versus brain research. But these explanations of fundamental human social instincts are *not* competing theories. They are intertwined (Fig. 1). We can study the shift from hunter-gatherer societies to city-dwelling, where the very necessity of permanence required and reinforced social practices that, for all practical purposes, became permanent as well—they became practices laid down in the social record that we continue to unearth.

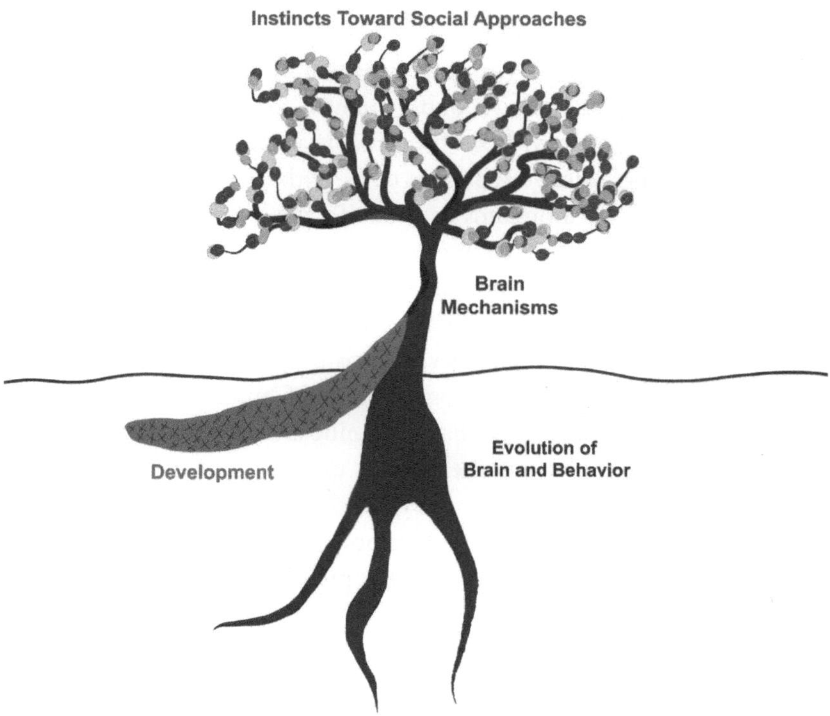

Fig. 1 Three intertwined causal routes help to explain the human instinct for social approach and contact: evolution, development and brain mechanisms. What evolves are the developmental capacities (programs) to produce the brain mechanisms for such human social instincts.

Simultaneously, what evolved tens of thousands of years ago are, indeed, the mechanisms necessary for a baby to develop from needs-based social reflexes to advanced capacities for social recognition and affiliation.

In turn, these advanced behavioral developments depend on the brain mechanisms neuroscientists study.

Thus, the causal routes to our sociality build on each other. At points in the process, one will predominate over the other—at one point, neuroscience may account for most of the explanation while, at other times, evolution or on a baby's development may be the most important focus of study. The point is that when viewed over the scale of *Homo sapiens*' development, no one explanation is sufficient. All three operate over eons, and I try to take account of each relative to the other.

Time scales

Because this book is kaleidoscopic in its approach—merging together very disparate enquiries and disciplines—the time-scales represented are disparate as well. The book deals with brain mechanisms whose time scales are measured in thousandths of a second. Its coverage of human development of social capacities deals with months and years. The next chapter's discussion of anthropological discoveries deals with phenomena which took tens, even hundreds of thousands of years. To the extent that I discuss the non-primate record, we are talking hundreds of millions of years. Spatially, my discussion is also scaled: it ranges from the minutest structures (e.g., genes) to large communities (e.g., modern cities).

This book emphasizes brain mechanisms because their study offers the newest perspective on the physiological bases of human society. However, all three intertwined approaches that I examine are currently dynamic fields of scholarship. In combination, the effect is uniquely dynamic. The effect is kaleidoscopic because social issue merge with and bear on evolution, brain mechanisms, and babies' childhood development. But I investigate each approach separately, showing how no one approach is sufficient without the others.

Coronaviruses

During the novel coronavirus pandemic, people around the world were instructed to adhere to "social distancing." This was hard. In effect, we were told to contravene everything that this book says is natural. But human beings are creative, and profoundly motivated to breach the distances between them. So, we used unnatural means to achieve natural ends: FaceTime, Facebook, Zoom, Skype, WeChat, Instagram, TikTok, GoToMeeting, Slack, Trello, Microsoft Teams, Confluence, Samepage, Free Conference Call. We talked, viewed each other, and collaborated virtually. A whole new category of pajama jokes developed, as we sprang on these platforms spontaneously and *as if* they were natural—everyone, presumably, got in the spirit. The desire to remain connected, and the strategies that we pursued, turned us into cyborgs-in-pajamas for a while. But what mattered was the connection.

As *Origins* proceeds, however, I will raise questions about whether our ancient brain mechanisms—developed at a time when people lived in tiny communities—are compatible with the infinite communities available online. I will ask whether technologies, naturalized over the short term to *feel* natural, are actually disorienting in the longer term. The coronavirus has only intensified trends that have been accelerating for the past 20 years, and I will ask whether the resulting co-evolution of humans and digital platforms (and the dependence of humans on these platforms for what simulates intimacy) could push up against the brain's information processing limitations.

Historical perspectives

In effect, *Origins* asks (and attempts to answer) a pair of venerable questions: why does society exist and how did it emerge? The average person might say "Well, obviously, I am drawn to other people because they help me and keep me company. That's been true since there *were* people." Well, sort of. Such a tautological approach (I am drawn to people because I am drawn to people) is woefully simplistic. It elides most of evolution and hundreds of years of thought—not to mention the latest neuroscience. It fails to recognize that the origin of society has been contested on ideological grounds, with various theories often reflecting the politics of when they were proposed. Such theories were highly speculative and, to the extent that they captured some truism about human nature, they were still not empirical.

For example, in the 17th century Thomas Hobbes argued that society emerged only with the founding of a state, which provided security in return for the surrender of some part of each individuals' freedom. The state brought people together. The extraordinary frontispiece of Hobbes' great book, *Leviathan*, showed the body of a crowned figure consisting of hundreds of faces—his subjects. The book appeared (1651) during the

Commonwealth period in England when the monarchy had been deposed, and it clearly favors the monarch's return (which occurred in 1660). Across the centuries, *Leviathan* has been a favorite of conservative thinkers. Some modern anthropologists, following its line of reasoning, suggest that hunter-gatherers lacked the institutions (law, organized religion, large-scale cooperation and food production) that made (and make) "society" possible. Cognitively and behaviorally, hunter-gatherers were loners until the imposition of authority. The idea is that even while the potential for society existed, society did not always exist.

In the 18th century, Jean-Jacques Rousseau argued that a "social contract" made society possible, so that a moral and collective body replaced the individual personality: "Each of us puts his person ad all his power in common under the supreme direction of the general will, and, in our corporate capacity, we receive each member as an indivisible part of the whole." Like Hobbes, Rousseau saw the need for sovereignty, but unlike him he vested it in an entire population. This was a radical departure from established norms, setting off revisionist thinking that culminated in the French Revolution. It was premised on social transformation from a "state of nature" (defined by self-interest) to a type of rational society (defined by a shared morality):

> *The passage from the state of nature to the civil state produces a very remarkable change in man, by substituting justice for instinct in his conduct, and giving his actions the morality [that] they formally lacked. Then only, when the voice of duty takes the place of physical impulses and right of appetite, does man, who so far had considered only himself, find that he is forced to act on different principles, and to consult his reason before listening to his inclinations.*

For Rousseau, society's emergence coincided with a commitment to shared—rather than a default adoption of individualized—welfare. People coalesced around the perception that collectivity was better (for each of them) than reliance upon each person's "physical impulses and right of appetite." Such impulses did not disappear, but were effectively suppressed as a new type of "interest" replaced self-interest. Rousseau had immense political influence, not least on the early 19th century German Romantics, in part, perhaps, because his theories were not empirically based.

During the 19th century, an array of social thinkers proposed arcane theories that posited various units of social organization. Lewis Henry Morgan, for example, proposed the "gens," or what we would call the clan. The clan was kinship-based, and begins to sound like modern anthropology, although it does not explain modern social organization. A great parade of thinkers followed, including Emile Durkheim and Sigmund Freud, both of whom focused on the importance to society of observing certain sexual norms.

In the 1960s, Marshall Sahlins argued that following the Stone Age, "indiscriminate sexuality, dominance, and brute competition" were overcome or subordinated, thereby securing "the evolutionary future of the species." But in effect, this argument just assumes the conclusion—how, after all, did early culture bring off this early confrontation with brute human nature? What made hunter-gatherers decide to substitute solidarity over conflict? To me, Sahlins sounds like he was projecting backwards from an ethos of Flower Power. Later in the century and into the 21st, feminist arguments about society's origins become prominent, but the landscape varies: while there are claims for women's crucial role in domesticating males (see Chris Knight, Sarah Hrdy), there are also more traditional views of male dominance.

In the following chapters, I will look at several theorists, many of whom have informed my own work. My point here has only been to illustrate the vast number of approaches to the question of why humans so deeply engage each other, and to suggest the current need for a scientific approach that draws on several areas of science to create a unified, nuanced, and appropriately complex theory of prosocial behavior. Any such approach was impossible until recently, since prior theorists lacked our current knowledge and, I must add, were more or less siloed within their disciplines.

I should add that this is not in the least a sort of inside-baseball quest to answer a question that doesn't much matter (except, maybe, to zealots in the disciplines that I draw on). If we do not understand how the brain works to promote prosocial action, we cannot understand why it fails—for example, in autism, where affected individuals turn in on themselves. Nor can we understand why some people (mainly men) are overly aggressive. We cannot understand the effects on the brain of loneliness. And we certainly cannot understand whether our immersion in social media may ultimately be unhealthy. *Social*, therefore, attempts to lay the necessary groundwork so that we can approach these questions with information and empathy.

While this book builds on what seems obvious, it explains *why* human connection *is* so obvious. We take society for granted, but I want to look under the hood. While I will range across scientific disciplines, I will emphasize brain mechanisms. This is because such mechanisms are immensely influential, and explain behavioral tendencies of which we are normally unaware but that drive our explicit behavioral patterns.

While all human beings have the desire to connect, my own perspective—which posits the inevitability of connection—may depend on living in a big city. Since most people now live in cities, that personal perspective is not necessarily a limitation (it simply coincides what must be the prevailing perceived reality). Nevertheless, I am going for more in this book: a look at the social spark born long ago in human history, in the baby's first days, and in multiple areas of the human brain.

I first got the idea to write about the social instinct by living in New York City, where 8.6 million people come in contact every day. I am inspired by New York City stories that describe the intensity and endless variation of those contacts. (Hamill, 2004; Wallace, 2017). But almost anywhere you go in the world, it's the same—a constant bustle that is only possible because people connect and reconnect kaleidoscopically. Billions of generous, prosocial acts occur every day. Some acts not so nice but, in every instance, somebody acts *toward* somebody else. What is the basis for these acts, apart from the immediate nature of the transaction? What in the evolutionary arc has made them possible, replicated on a smaller scale in a baby's social development and then, minutely, in the brain?

The goal of this book will be to understand the very first sparks of human instinct toward social approach. On a subway in New York City, a smile when someone cedes their seat… a shove when the car is overcrowded. In the airport in Guangzhou, help with a small child or a pat-down by Security. All of these and trillions of other examples show a human propensity to move *toward* another human, instinctively or after considered thought.

Population densities can be incredibly high (19,300/sq. mile in New Delhi). Each year, more people leave rural area for cities. There is no choice but to deal with each other. Behavior of one individual toward another, "social behavior," is a necessity. This book seeks to explain the initiation of the trillions of contacts which happen every day, not the valences.

Author's current perspective

For centuries, thinkers have argued about the foundations of human behavior in society. Now, scientific answers are becoming available. Anthropologists point to transitions from hunter/gathers to urban dwellers and to processes of domestication. Developmental psychologists can highlight amazing social competences in little babies. Neuroscientists currently implicate specific genetic and neurochemical mechanisms for social behavior, working through recently discovered neural circuits.

This book considers the phenomena of human affinity by combining these three disciplinary approaches that, in this context, have never been combined before: neuroscience (the study of brain mechanisms), paleoanthropology (evolution), and developmental psychology (babies). Its theory is that the first "spark" of human sociality—our initial, unreflective turn toward one another—is instinctive, governed by brain mechanisms that have evolved over eons, which are subject to babies' continued development. These three sets of explanations—anthropologic, developmental, neuroscientific—do not compete with each other. They are intertwined into a megafoundation for prosocial behaviors described here.

Now, in an age of "social distancing" consequent to Covid-19, our attention to social processes is heightened. I think the brain mechanisms highlighted here will foster the loving instincts that will bring us back together in post-Covid society.

In describing the origin of how we come to be "social" instinctively, I am describing a universal that is entirely physiological and not at all a feature limited to any particular culture. To use a common term, this movement of person A toward person B is the very opposite of a socially constructed phenomenon, variously interpreted according to spatial and temporal vectors. Someone on the steppes of Russia 10,000 years ago would recognize this movement because it is just a movement, a basic instinctive displacement of a body. Picture a baby reaching for its parent or a man reaching out to prevent another's fall. Right now we face profound social challenges deriving from Covid-19, economic collapse, and social unrest. The more that each reader appreciates that, as a species, we are designed for pro-social activity and have survived because of it, the more confident (s)he will be in the hard work required to rebuild. This book offers a *scientific rationale against despair*, however awful the situation now appears. Our "social behaviors" work almost all the time to spark, encourage and ensure civilized exchanges.

The book

This book provides three answers to this question of where the human social impulse comes from. As I indicated, each of these answers claims to explain on its own the human instinct to initiate social contacts:

First, our social brains *evolved*. Paleoanthropologists have deciphered the social capacities of certain non-human primates and proto-humans. Evolution works through changes in genetic capacities, so the genes involved in brain growth assume center stage.

Second, human babies must *develop* social awareness and social responsiveness.

Third, the *brain mechanisms extant* for human social behaviors did not pop up yesterday. There are vertebrate-wide neuronal groups, neuronal connections and genes expressed for which to account, present from fish to philosopher. For a full understanding of human social brain mechanisms, we need to sample certain non-human brains. In addition to those, special additions in visual and auditory systems increase the computational power of social circuits.

Again, these three approaches—evolutionary, developmental and mechanistic—should not be viewed as independent of each other, even though they may refer to separate scientific disciplines. Instead, all three are woven into a fabric of causal influences which, together, make us the social beings that we are.

The theory presented below illustrates friendliness as a special case of the most fundamental division among behaviors: approach versus avoidance.

Chapters

Evolution. *Chapter 1* traces the thinking of today's paleoanthropologists as they try to discern evolving patterns of social behaviors in protohumans. Larger social groups are correlated with bigger brains, and evolution toward bigger brains inevitably requires changes in gene expression. Therefore, this chapter introduces some of today's discoveries regarding genes associated with brain growth; these are associated with modern molecular and genetic techniques.

Babies on Social Trajectories. *Chapter 2*. Babies develop social skills early in life. Among typically developing humans, there are wide variations in various features of social behaviors: eye contacts, social recognition, joint attention, conversation. "Normal development" can be understood more by its product than its rate. The development of understanding with regard to another's intentions—so-called "theory of mind"—has been analyzed at the cortical level. I criticize imprecise thinking in this area.

Brains of humans and other vertebrates

Chapter 3. The *bauplan* (German for "structural plan"), the fundamental brain circuitry underlying social behavior, is actually laid down in fish brains and maintained throughout vertebrate evolution.

Chapter 4. Elemental social behaviors. Among various animals, the social behaviors which show up most reliably are sex, parental and aggressive behaviors. My own lab and others have accomplished hormonal and neuronal explanations of some of these behaviors.

Chapter 5. Visual social signals. Non-human primate brains have amazing circuits specialized for dealing with the sights of social interactions.

Chapter 6. Auditory social signals. From fish through frogs through birds and non-human primates sounds, including vocal signals, allow for social arrangements at a distance. Some scholars argue for a continuum of vocal signaling systems, rather than a sharp dichotomy from animal brains to humans.

The human (social) brain

Chapter 7. Beginning with British psychoanalyst John Bowlby, the central role of higher cognitive processes in our human social mentality has been studied extensively. This chapter will relate so-called "attachment theory" to studies to modern neuroscience.

Chapter 8. Altruism draws human beings together since humans naturally want to help others—even those whom they do not know personally (Fig. 2). Straightforward brain mechanisms can explain this impulse.

Fig. 2 A worker in New York City leaps onto the subway tracks to save another man from an onrushing train.

Chapter 9. Social avoidance. The most fundamental dichotomy between behaviors involves approaching a stimulus *as opposed to* avoiding it. For example, in neurobiological experiments with non-human primates looking at TV screens, the experimental tasks are often "Go/NoGo." Neuroscientist Helen Barbas and I think that social approach versus social avoidance decisions can be explained by a new and surprisingly straightforward theory of prefrontal cortex outputs through forebrain motor control regions.

Autism represents one set of conditions in which social aloofness prevents normal social behavior. Genes altered in autistic children have been analyzed extensively; gene regulation by epigenetic chemistries, as well. Boys are diagnosed about four times as frequently as girls, and I will explore nine explanations for this.

Chapter 10. Social loss. Another type of social loss is prolonged loneliness, which has real physiological consequences. It is amazing that people suffering from loneliness have reduced anti-inflammatory responses.

Chapter 11. Genetics and epigenetics. Gene expression is regulated in a manner that permits and even encourages some forms of social behavior. Hormones play a part, since they affect relevant brain circuits. "Epigenetics" refers to means by which gene expression is changed without changes in DNA itself. Damage to certain genes or to epigenetic mechanisms regulating those genes can damage social behavior.

Chapter 12. Overview and outlook. In the face of the planet's increasing population density and consequent, excessive demands on resources, our internal regulation of behavior toward others is more important than ever.

Because I am a neuroscientist, this book concentrates on brain mechanisms for the initiation of prosocial behaviors. But some would say that is wrong—that for explaining instincts for social connections we should look to the history of our human species: *paleoanthropology*. The first chapter of this book reviews what specialists in non-human primate social behavior and Neanderthals have concluded. Still another group of scholars emphasize the *developmental roots* of human social behaviors. The development of social eye gaze, parental recognition, and shared attention are all involved in this approach and are discussed in Chapter 2.

So, in determining the causes of human social approach—the instincts for human contact—is the issue brain mechanisms versus evolution of our species versus developmental psychology? No. As I have observed, these approaches should be combined. What evolved during non-human primate life and the lives of prehumen groups were precisely the brain mechanisms discussed here. Further, from the pre-term baby through the newborn through early social years, the brain mechanisms described here must develop through emerging temporal patterns of gene expression and neuroanatomical connections. The three roots of humans' instincts toward social contact are intertwined and, viewed as such, form a more comprehensive view of the human propensity toward social approach (Fig. 1).

Put another way, what evolves are the mechanisms. What develops are the mechanisms. What evolves are the capacities for that development. That is, mechanisms evolve with a species across time and then develop in each individual as they grow. So, while this book is pitched toward brain mechanisms, the evolutionary and developmental approaches are acknowledged through the first two chapters after this introduction.

Looking forward, we anticipate and we worry. On one hand, opportunities for certain types of social engagement through social media are rife. On the other hand, some biologists, psychologists and economists fear that our human brains are not equipped sufficiently to handle the magnitude and intensity of social contacts available first in Western countries and now worldwide. Our human brains do not scale to artificial group sizes accessible through Facebook and other social media.

If we are limited to "tribal brains," what will happen? The Outlook, with which I end the book, presents some informed views. I should note, however, that in view of the massive turn toward social media and collaborative platforms during the coronavirus pandemic, these questions take on an even greater urgency, since many psychologists, sociologists—and futurists—argue that we will never "revert" to forms of face-to-face communication that, only a short while ago, seemed to be the norm. To these theorists, we have entered a New Normal, where the size of each individual's community is potentially infinite. During the period of the coronavirus, we have become accustomed to this scaling up, accompanied by a kind of two-dimensional reality that somehow seems very easy to enter and sustain. Thus, while this book is about Evolution, it must necessarily examine whether—in some sense—humans have escaped the natural evolutionary process and taken it up in cyberspace, i.e., as a co-evolution with machines and social media platforms. Elements of such a co-evolutionary outlook begin to sound slightly scary. This is why understanding the bases of our natural social capacities is so crucial now, i.e., how far can we push them beyond human neurophysiology before the very idea of "social" becomes something completely different?

During the coronavirus pandemic, I realized the urgency of these questions when the manufacturer of my computer sent me an email: "Looking to stay connected? Upgrade your PC!" The company knew what was on everybody's minds, and naturally made the most of it. We *are* looking to stay connected, and we need the equipment. How we have always used our own "equipment" seems especially important to understand now.

References

Hamill, P., 2004. Downtown: My Manhattan. Little Brown, Boston, MA.
Wallace, M., 2017. Greater Gotham. Oxford University Press, Oxford.

Further reading

Brands, H.W., 2000. The First American: The Life and Times of Benjamin Franklin. Anchor, Random House, New York.

Chimpanzees do not build cities, nor do our other closest relatives among the great apes. What makes us different? Why are we the only urban apes?

From The Life and Death of Ancient Cities, (2020), page 20, by Greg Woolf.
(School of Advanced Studies at the University of London).

CHAPTER 1
Background: Evolution

Man is a singular creature. He has a set of gifts which make him unique among the animals....

(Bronowski, 1973, p. 19).

Humans possess an instinct to connect. We are curious about others; we form affinity groups; flirt, play sports, and start conversations spontaneously. All of these behaviors fall under the general rubric of creating "social" bonds, however temporary. Of course, to be social varies in intensity and from one context to another. But for purposes of this book, I simply mean that we have an instinctive behavior to seek and engage the company of others. So, the question is: where does this instinct come from? That is, where in the brain is it motivated, and how did this motivation develop over the course of our species' history?

As I state in the Introduction, there are three approaches to analyzing the origins of human social behavior: evolutionary (in this chapter); by reference to human development (next chapter), and by examining brain mechanisms (most of the book). I will concentrate on brain mechanisms, but will introduce them by first looking at evolution and human development. I start with evolution because it goes back the furthest. However, I want to repeat that while all three approaches will be analyzed separately, this is only for purposes of their presentation; ultimately, they come together kaleidoscopically. It is a fact of science that intellectual analysis needs neat compartmentalization—based on academic politics and the need to organize proliferating data—even though, in the real world, all these compartments easily open into each other.

Writing a book needs chapters and headings so that readers can follow the argument, but no systematization can adequately represent the criss-crossing, intersecting, and merged vectors that actually constitute long-term change. Thus, while I begin with evolution—the driving force in human development—I want readers to understand that the issue of how our species established social affinities must be seen as a multi-dimensional enquiry.

Origins of Human Socialization
https://doi.org/10.1016/B978-0-323-85861-8.00009-8

It is not a straight, one-dimensional procession from hunter-gatherers to tribes to urbanites living in an organized political entity. It is, rather, a story of increasingly large-brained hominids and their offspring's adaptation to communal life.

But before I begin to talk about evolution, I want to be clear about what, exactly, I think "evolved" to make social affinity possible. The ethologist Irenaus Eibl-Eibesfeldt (1989) observed that "animals congregate for different purposes": safety, sex, and especially care of the young. He observed that "the mother/offspring relationship is a reciprocal one" (p. 168), and that "the invention of parental care is the starting point for the development of higher differentiated social systems," e.g., friendly productive interactions, love, and the like (p. 169). As discussed in Chapter 9, these social relationships can be "marked by a distinct ambivalence" (p. 170)—do I approach? avoid? withdraw? After the initial approach occurs and there is a formation of trust, the major problem becomes group maintenance. *But this book is about the initial approach.* This chapter is about how that initial approach evolved. That is, before animals can congregate or start loving each other or even feel ambivalent about their communities, there has to be that first irresistible tug of A towards B and vice versa. I will examine how that tug arose and why, as a matter of evolution, it became a basic human motivation. Ultimately, of course (and in keeping with the real world's multidimensionality) this motivation became embedded in brain mechanisms and human children's development.

Of course, even narrowed down to this instinctive, initial tug, the word "social" can mean different things to different people—do you want to smile at someone, extend a helping hand, or feel a duty to join the National Guard? That is, one can imagine a spectrum of choices even within a narrow range. Without putting too fine a point on it, I simply mean that human individuals seek and engage the company of others—and I want to know where this impulse comes from and how it arose. This book assumes that humans possess a universal instinct to connect not limited to specific monetary, romantic, aggressive or other transactional situations. In this sense, the instinct is broad and undifferentiated, which suggests that it originated in the deep recesses of evolution before humans began living in specialized strata playing highly individualized roles. In other words, the instinct can be regarded as fundamental to our basic humanity.

In using the terms "basic," "universal," and "fundamental," I am excluding from my discussion all of the cultural epiphenomena surrounding our impulse to connect. Thus, I am not interested in marriage, even though it is a virtually universal practice. Nor am I interested in our tendency to form political parties or religious sects. I am not interested in conspiracies or secret societies like Skull & Bones or the Templars, Masons, or Knights of Columbus. Sororities and fraternities are not on my radar. All of these groups are projections of a desire to connect within an exclusive body that shares certain relationships and customs, and excludes individuals not permitted to share them. Humans love complex social formations where inclusion also strongly implies exclusion (even the marriage vow aspires to "forsaking all others"). But the impulse that I am addressing precedes its elaboration into culture. It is driven by neurophysiology. It evolved over eons of time vastly longer than the existence of any culture. It may be hard, in the 21st century, to envision a sort of bare humanity unconditioned by culture, although the coronavirus is forcing us in that direction. We are forced to contemplate our bare physicality, vulnerable to nature. I am writing about humans in that very elemental state.

I want to situate my discussion of evolution within a relatively narrow, specific compass: the evolution of brain mechanisms that account for our instinct to engage another human being. I say this up front because when we think of evolution we tend to think of whole species—how did birds evolve from dinosaurs, or dogs from wolves? But, in fact, evolution takes place at the micro-level and at the level of small structures. Darwin made a point of this when he began noticing the differentiation in certain birds' beaks. Chapter 7 of this book puts forward an amazingly straightforward neuroscientific theory of the origin of our instinct to connect based on the most recent neurobiological and genetic evidence.

Even in this brief sketch, I find it fascinating as a neuroscientist to explore the brain-based origins of human social behaviors from non-human primates and early hominids. Once evolved, the brain mechanisms which lie beneath modern, mature social behavior comprise a vertebrate-wide social brain: genes, neurons and circuits which support this instinct.

Social behavior mechanisms can be understood based on the fundamental division between social behaviors: "Should I approach or avoid?" This dichotomous choice depends on brain motor systems which use the concept of "embodied cognition" detailed in Chapter 9. Altogether I want to show how many genetic and epigenetic developments, operating through many causal routes, contribute to a human instinct towards prosocial behaviors. (By "epigenetic" I mean changes in the organism by a modification of gene expression rather than ("*epi*"-"above and beyond") alteration of the genetic code itself). The new CRISPR technique, which fundamentally alters the genetic DNA sequence, is the opposite of epigenetic change).

Concomitantly, the universe of neurobiological facts that help us understand human social behaviors allows us to understand the *absence* of such behaviors. Autism presents examples of social detachment which are of great current interest. But normal approach/avoidance behaviors and autism are two sides of the same coin insofar as we are trying to look at mechanisms that actually control the basic ability to connect. Of course, we have all heard of the autism "spectrum" and the related condition, Asperger's syndrome, in which individuals show some capability to extend themselves towards others while not seeming fully spontaneous or socially at ease. Understanding the basic neurobiology of human prosocial behavior helps to explain Asperger's as well, because we are talking about basic of mechanisms that can still go awry or function differently in different people. Variations among individuals—at the level of sociability—have a variety of explanations but, in a very real sense, they raise questions of epiphenomena. The basic brain mechanisms are still the starting point to understand the whole spectrum.

Before and during the evolution of *Homo sapiens* there were no experimental psychologists charting social behaviors in real time. So, writing about hominids (including early humans) can be controversial. But two subjects are not controversial. First, *domestication*—that is, staying in one place, adapting to the local environment, farming—offered pre- and early humans myriad opportunities for biologically adaptive social interactions. Second, humans have *big brains*. At the end of this background section, I will introduce some of the modern genetics analyzing brain growth during the evolution of our species.

Of course, for lower animals, prototypical social behaviors are sex behaviors, a well-studied subject (Pfaff, 2017; Pfaff and Joels, 2017). As a consequence of successful sexual contact, parental behaviors become necessary; these are complex (Numan and Insel, 2003). Sometimes, as classically described, nurturing parental behaviors are offered by females other than the mother (Hrdy, 2009; Chodorow, 1978). But beyond sex and parental behaviors, the vast array of all friendly behavior lies ready to explain. If, as Cambridge professor Barry Keverne's argues, positive social behaviors become "emancipated" from hormones that affect the brain, then all kinds of non-sexual, non-parental prosocial behaviors can emerge.

Domestication. Beyond these elementary sex-based behaviors, we can chart increasing capacities for social involvement citing studies of nonhuman primates and then Neanderthals. While there is no proof of an absolutely smooth, continuous development, what follows is a thumbnail sketch of evolving social capacities, featuring work of some of the most perceptive scholars. The key word emerging from this reading is *domestication*. "Domestication" is a process that results in individuals' living in accordance with norms, i.e., recurring practices that are suited to the individuals and their survival in place. Just as a wild anima becomes domesticated when it learns that its master will provide food but demand certain basic hygiene, so hominids achieved a level of domestication—over time—as they settled into what at first were rudimentary groups whose purpose was mutual protection. Domestication frequently entails the adoption of roles, e.g., according to gender, and even the establishment of hierarchies. Domestication depends on individuals' ability to feel, desire, and place a value on social interaction. Domestication is evidence for this experience, and provides the means to reinforce it.

Primates. With regard to social behaviors in non-human primates, some studies emphasize the positive and prosocial. For example, de Waal and Suchak (2010) review "well-developed prosocial and cooperative tendencies as reflected in the way primates support each other in fights, hunt together, share food and console victims of aggression." Some such behaviors can be explained "only by assuming an altruistic impulse, which -- as in humans -- may be born from empathy with the recipient's need." I will discuss the role of human altruism/empathy later in this book, but it is extraordinary, and supportive of my argument, to find such impulses attributed to primates lower down the evolutionary chain. It suggests that survival, even in relatively primitive groups, requires a sense of shared concerns, identification with one's fellows, and the importance of acting out certain bedrock impulses to affiliate. It suggests that even within such groups, there are practices that underlie—perhaps entail—the development of reciprocity, i.e., an ongoing give-and-take that make domesticity (survival over time in a given habitat) a greater likelihood in the face of continuing threats, scarcity, and losses.

Affiliative behaviors in wild animals have obvious beneficial effects, especially on the survival of infants (Silk et al., 2003; Archie et al., 2014). I presented a parsimonious theory of the mechanisms underlying the related altruistic behavior (Pfaff, 2014). Clay and de Waal (2013) review evidence from apes which "has shown that, after social conflicts, bystanders approach victims of aggression to offer stress-alleviating contact behavior, a phenomenon known as consolation." Different forms of empathic behavior with predominantly cognitive or emotional features have been distinguished by de Waal and Preston (2017). I am struck by the degree of behavioral differentiation—protecting infants, offering consolation—suggesting that the basic instinct has been adapted to the animal's environment over time. It is there, no matter what, because, over time, a great many matters will have arisen to prompt it. It kicks in to protect the animal and create bonds of reciprocity because the species has displayed and relied on the instinct as it has evolved.

Now consider a different kind of prosocial behavior: chimpanzees, sometimes said to be our closest non-human relatives, have prolonged mother-infant interactions (Hayashi and Matsuzawa, 2017). Chimps and humans have "a common grandmother" (Harari, 2014) and have DNA sequences which differ by less than 2%. There are descriptions of their "rearing of a disabled infant by a multiparous but inexperienced chimpanzee, and the carrying of dead infants in wild chimpanzees" thus revealing a strong affectionate bond between mothers and infants. Matsuzawa (2013) also points out that chimpanzees "apparently focus on the salient objects, neglecting the social context, while humans always recognize things within the social context." However, chimps do show lots of socially learned play.

Seyfarth and Cheney (2011) and Cheney et al. (2015) extend the discussion of friendly behaviors to males. Male baboons befriend lactating females and, in this regard, take physical risks reflected in high stress hormone levels. Helping with paternal behavior, they reduce their mating opportunities, as well. But apparently the altruistic, affiliative impulse is sufficiently strong to overcome the instinct to mate and, perhaps, to avoid situations that engender stress. Here, the baboon is not acting in his own interest, i.e., in the interest of individual survival, but because he is disinterestedly impelled to offer help to a member of his species.

It is easy to see how these prosocial behaviors evolved because friendships among non-human primates are adaptive to a group's environment, i.e., they are part of the process of domestication. They are promoted by genetic relatedness (individuals tend to help others who resemble them); more frequent among females than males (largely because of their role in rearing the young); and may result from a history of cooperative interactions that reinforces the practices (Seyfarth and Cheney, 2011). In this latter regard, individuals are capable of learning that if they follow a prosocial impulse there will be some benefit—either immediately or later on a reciprocal basis. If there is no perceived benefit, as in the baboon who helps the lactating female, there is still some likely inherited need which demands action towards the other.

Other researchers emphasize unfriendly social behavior. In fact, nonhuman primates do offer examples of extremely aggressive social behaviors (Wrangham and Peterson, 1996). Richard Wrangham refers to chimpanzees: "they were a good fighting party, seven males ..." (p. 5). He discusses a male "becoming enraged" and beating up his sister because she would not mate with him. Males may rape females with the aim achieving social control, "rather than fertilization in the present" (p. 141). Male chimpanzees "defend their territories ferociously" (p. 130). Wrangham's description of a defeated male "bitten and bruised and torn" and finally dead, is gruesome.

In *The Neanderthal's Necklace: In search of the first thinkers* (2001), Asuaga notes that our DNA can be distinguished from the chimp's by only about 1.6%. As we begin to sketch a brief background on hominid social behaviors, it is clear from the studies just quoted that these behaviors were not evolved out of nothing, but from elemental behavior patterns accompanied by strong positive or negative social valences. As I observed above, pro- and antisocial behaviors are two sides of the same coin, with various epiphenomena accounting for why an essentially prosocial brain mechanism can pitch into reverse. In all cases, it is the underlying mechanism that accounts for the behavior, albeit one or another situation can result in an antisocial outcome.

Neanderthals. Neanderthals and Cro-Magnons were hunter-gatherers. Obviously, there is a severe limit with regard to how many people can constitute a hunter/gatherer group in order to sustain the relative density of animal and plant life. Anthropologists and archeologists argue over Neanderthals' cognitive capacities, but they did make tools, which implies social transmission of tool-making knowledge. Their adornment, revealed in their burials, implies that in life the adorned individual sought others' social approval (it could also have been evidence of an ego that accompanies enhanced status within the group). Anthropologists guess that Neanderthals had vocal canals and ears adequate for producing and responding to speech. To the extent that there was speech, however primitive, the implication is that there were (social) conversations involving matters as mundane as cutting up an animal to more sophisticated concerns like choosing an experienced leader for a long-distance hunt. If you stop to think about it, even though we talk about the "transition" from hunter-gatherer existence to more complex forms of social organization, *even* hunter-gatherers had to call upon each other and *must* have developed a group dynamic adapted to their lives in the wild (Fig. 1). Just obtaining a woman to mate with required some sort of give-and-take—unless every mating was tantamount to rape.

Fig. 1 In human prehistory, a major transition from hunter/gatherer existence to living in crowded cities.

Thus, we have to assume that groups of Neanderthals had developed social signals and more extensive interactions as part of meeting the challenges to their survival.

As I have indicated and now want to emphasize, survival is key to understanding any evolutionary argument. Evolution favors adaptations that permit the species to survive. It is indifferent to individuals, and is driven by how well the group—i.e., the species—can accommodate to its environment. Sociality reinforces the group. It fosters mutual reliance, which is way stronger than self-reliance out in the wild. Thus, when we hear about American self-reliance, we have to recognize it as a cultural practice that is inconsistent with how the species evolved. We tend to think of self-reliance as a survival strategy, a sort of all-American, Davy Crockett version of manliness. But this is not what evolution favors. We might feel less inclined to rely solely on our own resources if we realized that we can do so only to the extent that the culture makes that possible. Even Davy Crockett had a gun, an artifact of 19th century industry. When Henry David Thoreau wrote *Walden* (1854), living alone on a cabin on Walden Pond, he still went into town occasionally. He had to. Likewise, Ralph Waldo Emerson's famous essay, "Self-Reliance" (1841), is not about backing away from others. Rather, it counseled non-conformity and a kind of radical individualism, but did not suggest that we shun cooperation to go it alone. Individuality of this sort entirely is consistent with collaboration of the social instinct. For the species, instinct is actually better for us than any upstart cultural claims to the contrary.

Kay Pruefer and her team (2017) at the Max Planck Institute for Evolutionary Anthropology in Leipzig, achieved DNA sequencing of a female Neanderthal who lived in Vindija cave in Croatia about 50,000 years ago. From the results, she inferred that nearby Neanderthal populations would have been of small size and that near relatives likely mated with ancestors of humans. From the Neanderthal data available, she estimates that roughly 2% of modern human DNA comes from Neanderthal genomes. Martin Sikora and his team (2017) at the Centre for Geogenetics in Copenhagen worked with a somewhat more recent site in western Russia, dating from about 34,000 years ago. Based on complex statistical analyses they suggest that the hunter-gatherer groups were small and to some extent inbred. There was apparently more inbreeding in a previous Neanderthal group from an earlier period in southern Siberia.

Of course, extending these and related findings, like patterns of co-burial, to inferences about social behavior involves guesswork. However, it looks as though groups only distantly related to each other overlapped socially by virtue of population movements. This inter-group mating would have required social attraction, signaling, and interaction at a level sufficient to deter any potential opposition or retaliation. It would have required a growing acceptance over generations as groups met and mingled, perhaps on land that could plausibly have been claimed by either or both groups. It could have engendered cooperation and primitive cultural exchange. My point is that there is evidence, going way back in hominid history, for social impulses—and hence the necessary brain mechanisms—that could even enable mating among possibly rivalrous groups. There might have been mating even if these groups still competed in other respects, e.g., for resources. To cooperate and compete simultaneously requires an even higher degree of social capacity so that one activity can proceed while another is subject to contestation. We can only conclude that hominid brain mechanisms for social interaction evolved to meet complex challenges—perhaps more complex than we give Neanderthals credit for.

I should note that long before this 30- to 50-thousand-year time period, *Homo sapiens* were making advanced tools in Kenya, Morocco, and Ethiopia, and enjoying the benefits of expanded social networks. Tool-making, like reinventing the wheel, is not something that every generation normally undertakes. Rather, it entails the inter-generational passage of knowledge which, in turn, entails social interaction on a rather sophisticated scale. It requires teachers and demonstrations. It requires practice and approval and, most likely, the sharing of tools among group members. The presence of tools, even if they are rudimentary, provides evidence of a prosocial impulse to teach, supervise, and ultimately create on an enhanced inter-group scale. It suggests that brain mechanisms had evolved to enable this impulse.

Domestication, again. Paleoanthropologists working in Turkey have documented evidence pointing to a shift from low-density hunter-gatherer populations to crowded towns. As I have already observed, there had to be upper limits on the number of hunter-gatherers in any one place in order to maintain a sufficiently high ratio between hominids and prey (and, indeed, hominids and plants which were eaten by hominids as well as their prey). The upside of low population density, however, was that the demands on one's social skills—though hardly non-existent—were fewer and simpler. The available space made the need for social skills literally less pressing. On the other hand, individuals in towns, who followed an impulse to live at close quarters, must have possessed considerable social skills. In proto-urban environments, one has less control over one's movements—and, indeed, over life in general—because one's actions impinge on others; one has to literally negotiate one's passage through crowded areas and through one's daily tasks. There is no room for the haphazard encounter. Rather, one's passage is narrowly defined, i.e., designated in advance, so that encountering others on the same defined route and in the same setting becomes inevitable. One has to interact with those others, presumably in accordance with an attraction/avoidance protocol that is common to the population. In other words, to live an urbanized life means that there is little room for passing by others unnoticed; everyone needs to share a capacity for effective social signaling and exchanges.

Cultural movement towards domestication can also be traced indirectly through the genetics of horses as their uses changed. Gaunitz et al. (2018) reconstructed the genomes of 42 ancient horses, domesticated about 4000 years ago, from remains at archeological sites. Evidence of domestication at these sites included signs of leather making and dairy production in sedentary settlements, in contrast to bands that survived through hunting expeditions. The authors were concerned to show these horses' genetic relations to their feral background—horse husbandry began only about 5500 years ago—but also to explain the transition from fitness for hunting to exploitation of horse milk, tolerance for harnessing and other domestic requirements. Of course, the need to raise horses, and the inclination to trade and race them, would have required an immense advance in social capability, not least because of the cultural institutions that societies would need to facilitate these activities.

So, if we are looking for an historic period when the social *instinct* gained traction, this would seem to be the expanse of time (varying among continents and cultures) during which hunters-gatherers settled down and became domesticated. Such periods would have offered the opportunity to develop socialization skills and practices. Scholars discussed below have considered development during this time period in as much detail as the data permits.

Despite great uncertainty about pre-human habits and capacities, some scientists have attempted to explain (in broad outline) what occurred. Geneticist David Reich (2018) discusses the beginning of the transition about 44,000 years ago from "cave men," i.e., Neanderthals, to modern *Homo sapiens*. "Within a few thousand years, Neanderthals were extinct. Did they have sex with each other, Neanderthal with human? Yes, but not much. Modern humans outside Africa have about 2% 'Neanderthal DNA.'" That Neanderthal DNA could not have been successful in the sense that it did not propagate; this is inferred because it is not greater than in peoples far from the region, Chinese and New Guineans. It did not spread.

In *The Humans who went Extinct* (2009), Denis Finlayson speaks of transitions from the lives of hunters to farmers', and then the "evolution of civilized societies out of an agricultural template" (p. 200). Hunter-gatherers of the steppe-tundra had "produced surpluses, and the need to manage those surpluses required more sedentary lives, consistent with (though not requiring) agricultural lives." "Structured societies with all the paraphernalia of civilization" could arise with or without agriculture (p. 201). In Finlayson's view, over thousands of years "domestication was not a complete break with the past but rather part of a continuum of increasing human intervention from predation to genetic engineering" (p. 203). Domesticated humans, accompanied by domesticated plants and animals, were protected in ways that allowed them to "out-compete their wild ancestors" (p. 204). By "out-compete," Finlayson means an ability to manipulate the environment (and the resources that it provides) so that it was easier to support populations. His reference to genetic engineering is to breeding techniques which, of course, are easier to employ where plants and animals are stationary. To accomplish these feats, individuals needed to have the impulse to congregate, to share close quarters and—essentially—to share. Their brain mechanisms had to be sufficiently developed to promote such impulses.

Pat Shipman shrinks from drawing hard and fast lines between Neanderthals and groups which would evolve into modern humans (Shipman, 2015). Perhaps most important in this regard are those long stretches of modern human DNA devoid of Neanderthal DNA sequences: "Neanderthal gene deserts" (p. 28). The capacity for the modern human cell nucleus to transcribe Neanderthal gene alleles which are linked to various diseases would be predicted to be very low. So, we likely are safe from Neanderthal-type diseases.

Two of Shipman's major themes involve "invasion biology." One of them is "how humans and their dogs drove Neanderthals to extinction" (Fig. 2). The story concerns domestication, i.e., the domestication of wolves that ultimately produced dogs. Shipman believes that modern dogs originated in Europe, because the largest samples of early dog DNA, between 18,000 and 32,000 years ago, are found there. During that period, wolf-dogs, natural hunters, helped humans kill mammoths, which were huge sources of meat. Wolf-dogs retained wolves' abilities to live in groups and hunt in packs. As such, their cooperative social hunting, as domesticates, might have provided a template for cooperative hunting by humans (p. 197). But in any case, humans had to band together to control these animal packs. They had to have brain mechanisms that encouraged such prosocial behavior.

With respect to domestication, archeologists Emily Hammer and Ben Arbuckle, as recounted in *Science* by Andrew Lawler, seek to explain how urban life arose. In their account, "farming started in the Near East about 10,000 years ago," but they ask whether those early humans had to travel, season-by-season, to follow the best farming weather. It has also been inferred that such nomads—if they were nomads—brought their livestock

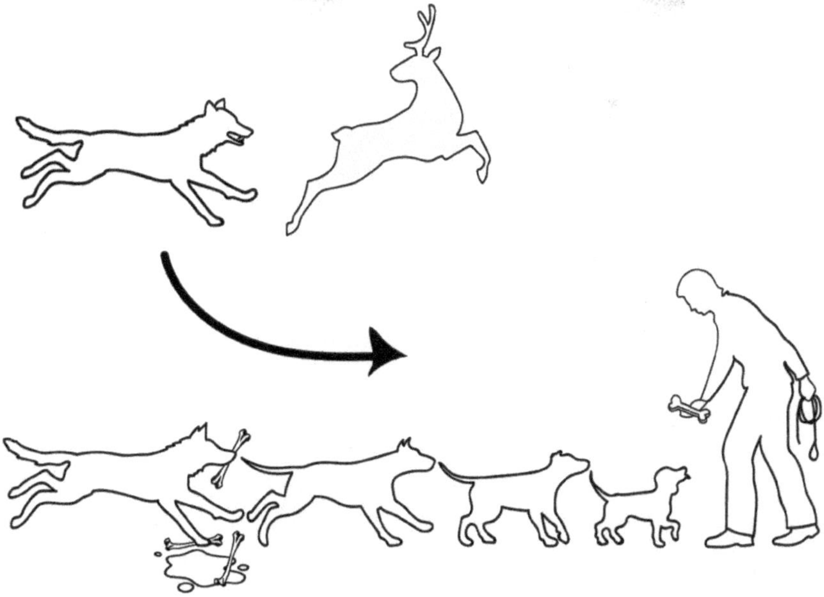

Fig. 2 Important in early human history: Domestication of plants and animals, illustrated here by wolves-to-dogs.

even while they pursued a trade of "valuable stones, metals and timber." These were not long-distance trades, but the farmers met "urban demands for meat and milk, as well as the wool that helped drive the Mesopotamian textile industry." In exchange, they got commodities that they could use or resell. What made it all possible were brain mechanisms, then already well-developed, that promoted social instincts. As I have already said, phenomena associated with sociality could not have occurred but for these brain mechanisms. In this sense, domestication does not present a chicken-and-egg question. We understand the order of things: without the brain mechanisms that promote social impulses, and encourage people to seek out and rely on other people, the phenomenon of socialization would not have occurred. Evolution is the handmaiden of the brain mechanisms, i.e., they had to evolve. But until they did, social formations that relied on relatively stationary human groups would not have been possible.

But there has been some significant work on sociability that goes beyond relatively recent domestication. Brooks et al. (2018) described evidence of complex symbolism and technology which suggested behaviors typical of *Homo sapiens* as early as 200,000 years ago. But these authors wanted more direct evidence of human behavioral origins. At Kenyan archeological sites they found evidence of long-distance stone transport from which "hominins made prepared cores and points, exploited iron-rich rocks to obtain red pigment, and procured stone tool materials." These activities reflected socioeconomic changes, the authors contend, relevant to human social and cognitive evolution.

A lot of theorizing about early socialization has to do with food. Richard Wrangham's 2004 book *Catching Fire; How Cooking Made Us Human* puts argues that cooking "changes our bodies, our brains, our use of time and our social lives." Cooking must follow the making of tools (e.g., for knives to dissect meat). The ability to soften tough meat "fostered human-like activities such as long-distance travel, big bodies, and increased cooperation." Cooking "makes our food safer, creates delicious tastes, reduces spoilage, allows us to soften and use tough foods." Cooking "increases the amount of energy our bodies obtain from our food" (p. 134), such that early humans survived and more successfully reproduced. As a result, genes whose expression facilitates these behaviors, were able to spread. For purposes of my argument, cooking is evidence of communality. Women cook together. Families consume cooked food together. There are social protocols for consuming it. If cooking changes out bodies and brains, our brain mechanisms make possible the rituals of cooking and consuming.

During the coronavirus pandemic, the *New York Times* and other media ran countless stories on how to maintain these rituals, recognizing their importance irrespective of immediate social constraints. Brain mechanisms continue to function even when individuals and families are cooped up at home without daily access to supermarket abundance. There was one story on all the things you can do with beans, and how "comforting" these homey dishes could be. The theme of the story was togetherness, uniting the family around its need for sustenance (a belated I-told-you-so, 200,000 years after the fact). Other stories talked about cooking with children, and making them part of the effort from food preparation to consumption. The point was for children to understand that the family was based, in large part, on mutual support and the consequent good feeling of being there for each other, i.e., we're in this together. Where individuals lived alone, stories suggested cooking together on Zoom, or some other social medium, to come as close as possible to preparing a meal with friends. Reading these stories as a neuroscientist, I picked up a subtext: the brain does not stop working just because everyone was working from home, sheltering in place, figuring out how to cope with a new normal. Brain mechanisms evolved to promote social bonding, and the evidence is everywhere.

Thus, from the earliest history of hominid, the capacity for social behavior has increased, until today we can hardly imagine not being social. Instagram and other social media have allowed us to make even our most private moments, like cuddling our kids before bedtime, a social event. We want to share our lives. "Sharing" is now a serious M.O. for millions of people disenchanted with capitalism. The "sharing economy" is an economic model defined as peer-to-peer (P2P) based activity for acquiring, providing, or sharing access to goods and service that is often facilitated by a community-based online platform. It is the opposite to conventional business-to-business (B2B) operations that are based on hiring employees and selling to consumers. Whether acknowledged or not, it recognizes ancient, brain-based social impulses. In fact, there are initiatives in this direction all over the place. Take, for example, Creative Commons, which allows works to be shared without the property restrictions of copyright. "Commons" is the key terms, and might be expanded to "Brain-Based Common Sense." How these brain mechanisms have evolved, eventually to serve modern human social behaviors, comprises the subject of this book.

The evolution of big brains

This next part of this chapter examines the genetic basis for the large-size human brain which, as research has shown, is linked to humans' prosocial impulses. But first I want to explain my train of thought, that is, how I became interested in human brain size as a product of and as a driver of such impulses.

My interest in the genes that foster brain growth was set in motion by results from Oxford professor Robin Dunbar (1992; Dunbar and Shultz, 2007) who showed that among non-human primates, size of the neocortex (relative to the rest of the brain) was correlated with the size of the animals' social groups. That is, the bigger their brains, the bigger the individuals' community—or, even more specifically, bigger brains are more prosocial. Dunbar speculated that it was "the particular demands of the more intense forms of pair bonding that was the critical factor that triggered this evolutionary development." That is, as individuals created more durable pairs in large communities, a feedback loop developed leading to brains that reinforced this social tendency.

Dunbar's results, however (which only involved non-human primates), did not settle all questions concerning the drivers of human brain-size evolution. Mauricio Gonzalez-Forero and Andy Gardner (2018; McElreath,

2018), in the School of Biology at Saint Andrews University in Scotland, used new computational analyses of the investments in and challenges to increased brain- and body-size (e.g., metabolic costs). Their theoretical model traded off ecological challenges (e.g., is there enough food for growth?) versus social challenges. Following a number of logical inferences and complicated calculations they stated that their "assessment fails to support social hypotheses as explanations for the evolution of human brain size, and is more consistent with ecological hypotheses" (p. 556).

Nevertheless, Chet Sherwood (2018), professor of anthropology at George Washington University, reminds us that not only do humans have large brains, but also that human brains show "disproportionate growth in cortical regions associated with high cognitive performance: the prefrontal cortex, temporal and parietal cortex regions." The Mirror Neuron system in the parietal cortex, crucial for the understanding of other peoples' intentions, is a good example of this (see Chapter 4) as is the oxytocin receptor system in the prefrontal cortex, involved in social behavior regulation (see Chapter 7) as well. These neurophysiological and neurochemical observations, consistent with Dunbar's thinking, suggest that increased brain size—in primates in general—correlate with an enhanced prosocial impulse.

There has also been substantial research linking a documented increase in IQ to the effects of increased population density. As a first step, consider that increasing IQs have been apparent for some time. Following up Dunbar's and Sherwood's results, and thinking about humans' cognitive performance in relation to brain size, Trahan et al. (2014), performed a rigorous meta-analysis that confirmed a rise in IQ scores over time (so-called "Flynn effect"), and highlighted the Flynn effect's "robustness across different age groups, measures, samples, and levels of performance." Flieller (1999) reported that the amount of gain observed over time "varied across tasks, being very large for combinatory thought but mixed for conservation. This acceleration of cognitive development can partially explain the continuous rise in intelligence test performance (Flynn effect)." The Flynn effect has been intensely analyzed (see Neisser, 1998). Now let's consider what causes the Flynn effect. Most reports, including some of Flynn's (1984, 1987, 1996, 1998), have looked at environmental causes. Prominent among those is urbanization, i.e., increased crowding. Even a short period of 25 years revealed the Flynn effect in China (Liu and Lynn, 2013) during a period of intense urbanization. Childhood urbanicity has been tied to variations in dopamine receptor genotypes and to prefrontal cortical function (Reed et al., 2018). Both of these measures are important to social approach, (see Chapter 7).

In Pietschnig andVoracek's (2015) meta-analysis of the Flynn effect, "favorable social multiplier effects" seem to be partly responsible for increases in IQ. Part of this must result from learning in a social setting—the trading of information, ideas, and techniques; the emulation of habits conducive to learning; and sheer competition for rewarded competence. (Indeed, in this era of individualized work-from home, it is acknowledged that such workers miss out on the stimulation afforded by a group setting). That is, social groups *encourage* teaching and assistance to benefit their members and, necessarily, the group as a whole. This feature ofbehavior in groups has been studied systematically. Dickens and Flynn (2001) emphasized social drivers, coining the phrase "social multiplier effects" and arguing that "social interaction can contribute to understanding large IQ differences between groups." (p. 347). In the tradition of Alison Jolly (1966), Kevin Laland (2017), professor of evolutionary biology at the University of Saint Andrews', emphasizes the importance of social learning related to brain volume: the ability to copy the new accomplishments of others, and the amount and accuracy of social learning (Street et al., 2017; Reader and Laland, 2002). The capacity to share multiplies human accomplishment and permits humans to populate large areas of the planet, sometimes with extremely high density.

Indeed, part of the creativity that is spurred by increased density involves the creation of technologies to enable the group's survival amidst increased density. Skyscrapers—which rely on elevators, hardened steel, and reinforced concrete—were not invented by nomads or even stay-put farmers. They (literally) arose in late 19th and early 20th cities: first in Chicago, then New York. The classic apartment buildings, office towers, and hotels of New York, some of which are in the National Register of Historic Places (along with the Brooklyn Bridge and Grand Central Terminal) are also of this vintage and could not have been feasible but for unprecedented numbers of people working and bustling about. Indeed, construction on this scale makes no sense but for the omnipresence of people.

Virus. During the coronavirus pandemic, the subways (which normally carry 4.3 million people a day!) seemed forlorn, almost imaginary. The airports seemed like abandoned movie sets. The point is that if you look around, you can see how communities create means to maintain connections among community members. It's obvious, indeed *more* obvious when the crowds are gone and we contemplate artifacts that once we took for granted. Moreover, while these means of connectivity are largely physical—like buildings, trains, and airports—and help to maintain physical connections, the newest technologies maintain instant, virtual connections that speed up connection but make it more abstract. The need for connection persists, but its nature evolves. What changes is the environment, while the brain keeps chugging.

In summary, among non-human primates, brain size is *correlated* with social demands for useful interpersonal activities including but not limited to sex. Among humans, average IQ has been creeping up, *correlated* with social crowding, i.e., urbanization. Switching to causal reasoning, I ask: what genes promote brain growth?

Evolution of genes for brain growth

It would be speculative to opine on specific brain mechanisms for social behaviors among the non-human primates described above. But we *can* focus on brain growth, correlated with social behaviors in ever-larger groups (Dunbar, 1992; Dunbar and Shultz, 2007). Here and in the chapters that follow, I will emphasize new work on genes causal to social behaviors and genes whose expression is correlated with social behaviors. I will link these to genes for brain growth. That is, I will discuss this sequence: genes drive brain growth which drives social behavior.

As we focus on genes related to brain growth, we keep in mind that in addition to simple measures of size, other factors are at least as important: the amount of folding and therefore cortical surface area, implying the amount of computing power resident in the underlying six layers of cortex. Further, given a gene's presence, its regulation in terms of the timing and strength of its expression are what molecular biologists must focus on now. Nevertheless, genes and their relation to brain size give us a reasonable topic to start thinking about.

Enard et al. (2002) wanted to study gene expression patterns in the human brain and the brains of non-human primates such as chimpanzees, orangutans and monkeys. Between 98% and 99% of DNA sequences were identical when humans and chimpanzees were compared. But these authors discovered that human patterns of gene expression (and protein expression, as studied in two-dimensional gels) in frontal cortex differ significantly from those of chimpanzees. And so, the application of chimpanzee brain findings to arguments about human brain evolution may be interesting but should be viewed skeptically.

Usually, when we read about DNA sequences, we think about the corresponding messenger RNA and protein sequences. However, in addition to mere expression, molecular biologists realize that they must focus on *regulation* of expression, as mentioned above. We now know that hierarchies of these regulatory processes are extremely complex, and that they include the production and action of tiny RNAs called "microRNAs."

It turns out that in prefrontal cortex and cerebellum between 10% and 30% of these microRNAs are different among humans, chimpanzees and monkeys. For example, the authors discovered a "signature of positive selection" in the upstream region of one miRNA with human-specific expression (Hu et al., 2011). Somel et al. (2011) agree about the importance of microRNAs. They calculated apparent rates of evolutionary change and concluded that humans' rates of change were faster than chimpanzees' or monkeys'. Some microRNAs might be much more important in this regard, and they proposed three particular microRNAs as key regulators.

Shulha et al. (2012) looked at a different kind of transcriptional regulatory change in DNA chemistry": DNA methylation, compared among human, chimpanzee, and monkey. Concentrating on transcriptional start sites (TSS), these authors studied prefrontal cortex and discovered 471 sequences with human-specific enrichment or depletion of DNA methylation. Therefore, coordinated epigenetic regulation via newly derived TSS-related chromatin could play an important role in the emergence of human-specific gene expression networks in the brain that contribute to cognitive functions and neurological disease susceptibility in modern humans.

Encouraged by their own and others' data, Somel et al. (2013) asked the ambitious question: "What evolutionary events led to the emergence of human cognition?" They decided to emphasize molecular events which would account for the very long developmental period over which human brains gain synapses. While they came up with several interesting genes whose sequences would separate the human from the Neanderthal and other pre-human lineages, I am reminded that RNA processing, protein synthetic mechanisms and, most important, higher-level transcriptional regulatory mechanisms (e.g., microRNAs and other non-coding RNAs) are likely to carry a share of the explanation.

Other molecular and genetic approaches to brain growth have been launched from stem cell biology using cerebral "organoids", three-dimensional nerve cell cultures intended to mimic actual brain structures. Thus, some molecular biologists can grow from stem cells three-dimensional structures that they hope will assume microanatomy reminiscent of some part of the brain. For example, the California stem cell biologist Melissa Lancaster (Lancaster et al., 2013) working in Austria, cultured neurons in such a manner as to "include a cerebral cortex containing progenitor populations that organize and produce mature cortical neuron subtypes." She hopes not only to recapitulate normal events in brain development but also to understand failures of normal development.

Further, Luo and Lancaster (Luo et al., 2016) compared patterns of gene expression and DNA methylation dynamics in their organoids and saw similarities to fetal human brain. Lancaster et al. (2017) improved their technique of culturing organoids by using microfilament polymers as "floating scaffolds" to improve the match to actual cortical development. Work in other labs focused on cells whose continued development will lead to nerve cells, i.e., "neural progenitors." Li et al. (2017) deleted a gene called PTEN, a tumor suppressor gene which acts as a regulator of the cell cycle. Deletion of PTEN caused production of more neural progenitors and, as a result, larger organoids. The authors hope that such a manipulation sheds light on how to make a larger and better folded cerebral cortex. All of this work seeks to understand exactly how our brains got as big as they are.

It must be said that compared to the delicacy and precision of normal developmental brain mechanisms, organoids look like blunt tools. The incredibly large number of normal developmental steps, tuned finely in space and time, make current organoid technology look awkward and impoverished. Organoid science is unlikely to achieve the exquisitely-regulated trajectories conceived by Tabansky et al. (2015) as leading to natural neural "cell types" defined by classes of functional equivalence. Thus, from organoids as they exist, we can learn only a few technical facts concerning brain growth.

All of this work has aimed at producing bigger brains, perhaps reminiscent, for some, of anthropologists' speculations about larger brains appearing during pre-hominid evolution. On the flip side, it is reasonable to ask whether today's genetics shed light on failures of brain growth, "microcephaly." The subject of brain failures will be difficult to sort out, because for every way that a brain can develop normally, there are thousands of ways for the process to go wrong. Nevertheless, we should consider some of what has been learned to date.

Mostly, the work has proceeded gene by gene. For example, Liao et al. (2013) tied mutations in the gene DPP6, which codes for a membrane protein involved in protein metabolism, to failure of full brain growth, "microcephaly" and mental retardation. Chen et al. (2014) performed similar analyses for a gene involved in mitosis, WDR62, mutations in which are associated with microcephaly apparently due to decreased neural progenitor cell numbers.

Larger numbers of genes are implicated in other neurodevelopmental disorders. Autism, featuring social, motor and cognitive disabilities, is now associated with more than 700 gene alterations. Some of those mutations may occur in continually dividing sperm, which can account for the "paternal age effect" (where older fathers tend to have children with a higher rate of defects). No single gene accounts for a large percentage of autism. Large numbers of the genes emphasized in the autism literature support synaptic functions. Currently, my former protégé Murat Kilinc, now at the Florida Scripps Research Institute, (Kilinc et al., 2018) highlights a gene called SYNGAP which regulates neurodevelopmental mechanisms and has significant behavioral phenotypes. Lomelino et al. (2017) brought into the conversation a gene whose product is required for synthesis of amino acids asparagine and glutamate, playing into normal protein synthesis and neurotransmitter function.

NOTCH is another gene which has long been of interest to neurobiologists for its role as a determinant of neuronal number in the mammalian cortex. Three forms of notch are in radial glia cells, crucial for developing neuronal migration and positioning. Different kinds of abnormalities of human-specific notch function "are associated with *macro*cephaly and autism; and (other) deletions with *micro*cephaly and schizophrenia" (Fiddes et al., 2018).

Not only the presence of a gene but also its degree of activity can influence changes important for the development of a human brain from a nonhuman primate brain. A gene called NOTCH2 is present in the brains of non-human primates and has long been implicated in brain development, but a new version called NOTCH2NL, though present in apes, seemed inactive. That could mean that it was not transcribed into mRNA or then not translated into protein. In humans, the NOTCH2NL human-specific paralog of the NOTCH2 receptor has become active (Fiddes et al., 2018; Suzuki et al., 2018) and promotes cortical expansion. As well, in the three-dimensional nerve cell structures made from stem cells called organoids, adding the new form of a NOTCH2 gene caused additional stem cell proliferation leading, eventually, to a bigger nerve cell organoid. As above, I note that while technically exciting, this discovery has limited implications for developmental neurobiology because in such organoids all the normal cell-to-cell signaling which guide normal brain structure formation are absent.

Turning to the big picture, we also can address questions of human diversity. Reardon et al. (2018) did magnetic resonance imaging studies using two large samples of human brains, one from Philadelphia hospitals and one from NIH, and discovered that larger human brains show greater areal expansion in frontoparietal cortical networks—regions responsible for higher cognitive activities—than in limbic, purely sensory or purely motor systems. The authors felt that this pattern of brain growth "recapitulates cortical remodeling across evolution." In particular, when investigating molecular correlates of this pattern of brain growth, mitochondrial genes stood out as well as related genes associated with a high degree of energy use: oxidative phosphorylation. Clearly, shortcomings in regulation of these genes would contribute to neurodevelopmental disorders. We can be certain that the numbers of genes discovered to be involved in such developmental problems will only increase.

In this chapter, we have begun to review one prong of my argument: evolution of the human brain in a way that favors the prosocial. Notice that I said "one" and not "the first" prong of my argument since—as I have explained—all three prongs are coordinate, i.e., intertwined strands of what, I believe, should be a complex multi-disciplinary approach to explaining the prosocial impulse in humans. Now notice as well that I said "impulse" since I am not explaining the vast array of behaviors that can be called prosocial. Thus, in this chapter, I sought to explain how humans evolved to respond to, i.e., to be attracted to another as their *default* response. There are a million reasons why, in any particular instance, we might avoid another person—but that is not our default response. Evolution has favored our interest in others, our interest in maintaining groups, because that interest promotes our survival. Through the eons, individuals have developed prosocial techniques enabling them to survive in groups and enabling groups to survive in the face of external challenges. Looking forward, when we examine the human brain and compare it to the brains of other hominids, we can see how our brains have evolved mechanisms that control and promote the prosocial. We will see how genes play a crucial role in our brains' development towards the prosocial. Over the course of evolution, humans have become physically adapted—not just simply practiced at—living in groups. We are not the only species so adapted, but we have evolved brain mechanisms that make us highly adapted.

Virus, again. We can see this adaptation on display when, all of a sudden, the natural means of remaining social are suddenly off-limits. If we cannot congregate, move around among groups, or even towards another individual we feel disoriented. We feel thrust into an unnatural situation that deprives us of a basic human right. During the coronavirus pandemic, pundits opined that the United States could never deprive people of the "right" to remain social in the manner of authoritarian regimes. This right was said to be Constitutional; much of the initial virus denialism reflected people's inability to imagine their lives except as part of a mobile society with infinite options. (Part of social media's appeal is that it replaces infinite physical mobility with an infinite array of virtual options). During the crisis, nobody said "Oh, my failure to observe 'social distancing' was just neuroscience in action." But to a degree, it was, compounded by selfishness and, perhaps, a sense of indestructibility. We were not just dealing with nature at the micro-level of the virus, but also at the level of brain mechanisms—which we can't exactly see but we still can understand.

In the next chapter, I will discuss the development of social behaviors early in children's lives. While the emergence of cognitive and emotional capacities delights all parents and grandparents, clinicians must be alert for departures from the normal developmental range that might signal a neurodevelopmental disorder.

Perspective and new questions

"*Homo sapiens* is primarily a social animal. Social cooperation is our key for survival and reproduction" (Harari, 2014, p. 26). From the earliest history of huminoids, capacities for social behaviors have increased in socially productive ways and by interesting causal routes. How brain mechanisms have evolved to serve modern social behaviors by humans, eventually, comprise the main subject of this book.

But questions remain.

1. How long will it take for us to understand the full panoply of genes required for normal brain growth?
2. Among those genes, are some susceptible to epigenetic regulation according to intrauterine experience?
3. But brain size is not everything—modern human brains actually are a bit smaller than you might have expected from the eons-long curve of human brain growth. Some Neanderthals had brains larger than modern humans. How did that come about? What does it mean?
4. Once social organization has been initiated, what forms does it take? Chiao (2010) talks about social status hierarchies as ubiquitous among animal species and highlights, among relevant mechanisms, "allelic variation in the serotonin transporter gene", as well as speculating about neural networks involved. In a convenient laboratory species, James Curley and his team at the University of Texas has analyzed the efficient formation of linear hierarchies with respect to their hormonal and neural mechanisms (Lee et al., 2019).

References

Archie, E.A., et al., 2014. Social affiliation matters: both same sex and opposite sex relationships predict survival in wild female baboons. Proc. R. Soc. B 281, 20141261.

Arsuaga, J.L., 2001. The Neanderthal's Necklace: In Search of the First Thinkers. Four Walls Eight Windows, New York.

Bronowski, J., 1973. The Ascent of Man. Little, Brown, Boston, MA.

Brooks, A.S., et al., 2018. Long-distance stone transport and pigment use in the earliest middle stone age. Science 360, 90–95.

Chen, J.F., et al., 2014. Microcephaly disease gene Wdr62 regulates mitotic progression of embryonic neural stem cells and brain size. Nat. Commun. 5, 3885–3889.

Cheney, D.L., et al., 2015. The costs of parental and mating effort for male baboons. Behav. Ecol. Sociobiol. 69 (2), 303–312.

Chiao, J., 2010. Neural basis of social status hierarchy across species. Curr. Opin. Neurobiol. 20, 803–809.

Chodorow, N., 1978. The Reproduction of Mothering. University of California Press, Berkeley.

Clay, Z., de Waal, F.B., 2013. Bonobos respond to distress in others: consolation across the age spectrum. PLoS One 8 (1), e55206. doi:10.1371.

de Waal, F.B.M., Preston, S.D., 2017. Mammalian empathy: behavioural manifestations and neural basis. Nat. Rev. Neurosci. 18 (8), 498–509.

de Waal, F.B.M., Suchak, M., 2010. Prosocial primates: selfish and unselfish motivations. Philos. Trans. R. Soc. Lond. Ser. B Biol. Sci. 365 (1553), 2711–2722.

Dickens, W.T., Flynn, J.R., 2001. Heritability estimates versus large environmental effects: the IQ paradox resolved. Psychol. Rev. 108, 346–369.

Dunbar, R.I.M., 1992. Neocortex size as a constraint on group size in primates. J. Human Evol. 20, 469–493.

Dunbar, R.I., Shultz, S., 2007. Evolution in the social brain. Science 317, 1344–1347.

Eibl-Eibesfeldt, I., 1989. Human Ethology. de Gruyter, New York.

Enard, W., et al., 2002. Intra-and interspecific variation in primate gene expression patterns. Science 296, 340–343.

Fiddes, I.T., et al., 2018. Human-specific $NOTCH_2NL$ genes affect notch signaling and cortical neurogenesis. Cell 173, 1356–1369.

Finlayson, D., 2009. The Humans Who Went Extinct. Oxford University Press, Oxford.

Flieller, A., 1999. Comparison of the development of formal thought in adolescent cohorts aged 10 to 15 years (1967-1996 and 1972-1993). Dev. Psychol. 35, 1018–1058.

Flynn, J.R., 1984. The mean IQ of Americans: massive gains 1932 to 1978. Psychol. Bull. 95, 29–51.

Flynn, J.R., 1987. Massive IQ gains in 14 nations: what IQ tests really measure. Psychol. Bull. 101, 171–191.

Flynn, J.R., 1996. What environmental factors affect intelligence: the relevance of IQ gains over time. In: Detterman, D.K. (Ed.), The Environment: Current Topics in Human Intelligence. vol. 5. Ablex, Norwood, NJ, pp. 17–29.

Flynn, J.R., 1998. IQ gains over time: toward finding the causes. In: Neisser, U. (Ed.), The Rising Curve: Long-Term Gains in IQ and Related Measures. American Psychological Association, Washington, DC, pp. 25–66.

Gaunitz, C., et al., 2018. Ancient genomes revisit the ancestry of domestic and Przewalski's horses. Science 360, 111–114.

Gonzalez-Forero, M., Gardner, A., 2018. Inference of ecological and social drivers of human brain-size evolution. Nature 557, 554–557.

Harari, Y.N., 2014. Sapiens: A Brief History of Humankind. Vintage, London.

Hayashi, M., Matsuzawa, T., 2017. Mother-infant interactions in captive and wild chimpanzees. Infant Behav. Dev. 48, 20–29.

Hrdy, S., 2009. Mothers and Others. MIT, The Belknap Press, Cambridge, MA.
Hu, H.Y., et al., 2011. MicroRNA expression and regulation in human, chimpanzee, and macaque brains. PLoS Genet. 7, e1002327.
Jolly, A., 1966. Lemur social behavior and primate intelligence. Science 153 (3735), 501–506.
Kilinc, M., et al., 2018. Species-conserved SYNGAP1 phenotypes associated with neurodevelopmental disorders. Mol. Cell. Neurosci. 91, 140–150.
Laland, K., 2017. Unfinished Symphony: How Culture Made the Human Mind. Princeton University Press, Princeton, NJ.
Lancaster, M.A., et al., 2013. Cerebral organoids model human brain development and microcephaly. Nature 501, 373–379.
Lancaster, M.A., et al., 2017. Guided self-organization and cortical plate formation in human brain organoids. Nat. Biotechnol. 35, 659–666.
Lee, W., et al., 2019. Hormones and behavior. Social status in mouse social hierarchies is associated with variation in oxytocin and vasopressin 1a receptor densities. Horm. Behav. 114, 104551.
Li, Y., et al., 2017. Induction of expansion and folding in human cerebral organoids. Cell Stem Cell 20, 385–396.
Liao, C., et al., 2013. Loss-of-function variation in the DPP6 gene is associated with autosomal dominant microcephaly and mental retardation. Eur. J. Med. Genet. 56, 484–489.
Liu, J., Lynn, R., 2013. An increase of intelligence in China 1986-2012. Intelligence 41 (5). https://doi.org/10.1016/j.intell.2013.06.017.
Lomelino, C.L., et al., 2017. Asparagine synthetase: function, structure, and role in disease. J. Biol. Chem. 292, 19952–19985.
Luo, C., et al., 2016. Cerebral organoids recapitulate epigenomic signatures of the human fetal brain. Cell Rep. 17, 3369–3384.
Matsuzawa, T., 2013. Evolution of the brain and social behavior in chimpanzees. Curr. Opin. Neurobiol. 23, 443–449.
McElreath, R., 2018. Sizing up human brain evolution. Nature 557, 496–499.
Neisser, U. (Ed.), 1998. The Rising Curve: Long-Term Gains in IQ and Related Measures. American Psychological Association, Washington, DC, pp. 125–154.
Numan, M., Insel, T., 2003. The Neurobiology of Parental Behavior. Springer, Heidelberg.
Pfaff, D.W., 2014. The Altruistic Brain: How we Are Naturally Good. Oxford University Press, New York.
Pfaff, D.W., 2017. How the Vertebrate Brain Regulates Behavior: Direct from the Lab. Harvard University Press, Cambridge, MA.
Pfaff, D.W., Joels, M. (Eds.), 2017. Hormones, Brain and Behavior, third ed. Elsevier, Cambridge (a 5 Volume Reference Treatise).
Pietschnig, J., Voracek, M., 2015. One century of global IQ gains: a formal meta-analysis of the Flynn effect (1909-2013). Perspect. Psychol. Sci. 10, 282–306.
Pruefer, K., et al., 2017. A high-coverage neandertal genome from Vindija cave in Croatia. Science 358, 655–658.
Reader, S.M., Laland, K.N., 2002. Social intelligence, innovation, and enhanced brain size in primates. Proc. Natl. Acad. Sci. U. S. A. 99, 4436–4441.
Reardon, P.K., et al., 2018. Normative brain size variations and brain shape diversity in humans. Science 360, 1222–1228.
Reed, J.L., et al., 2018. Interaction of childhood urbanicity and variation in dopamine genes alters adult prefrontal function as measured by functional magnetic resonance imaging (fMRI). PLoS One 13 (eCollection).
Reich, D., 2018. Who we Are and How we Got Here. Pantheon, New York.
Seyfarth, R.M., Cheney, D.L., 2011. The evolutionary origins of friendship. Annu. Rev. Psychol. 105, 35–83.
Sherwood, C.C., 2018. Professor of Anthropology at George Washington University. Scientific American, pp. 60–71.

Shipman, P., 2015. The Invaders. Harvard University Press, Cambridge, MA.
Shulha, H.P., et al., 2012. Human-specific histone methylation signatures at transcription start sites in prefrontal neurons. PLoS Biol. 10 (11), e1001427.
Sikora, M., et al., 2017. Ancient genomes show social and reproductive behavior of early upper Paleolithic foragers. Science 358, 659–662.
Silk, J.B., et al., 2003. Social bonds of female baboons enhance infant survival. Science 302, 1231–1234.
Somel, M., et al., 2011. MicroRNA-driven developmental remodeling in the brain distinguishes humans from other primates. PLoS Biol. 9 (12), e1001214.
Somel, M., et al., 2013. Human brain evolution: transcripts, metabolites and their regulators. Nat. Rev. Neurosci. 14 (2), 112–127.
Street, S.E., et al., 2017. Coevolution of cultural intelligence, extended life history, sociality, and brain size in primates. Proc. Natl. Acad. Sci. U. S. A. pii: 201620734.
Suzuki, I.K., et al., 2018. Human-specific $NOTCH_2NL$ genes expand cortical neurogenesis through Delta/Notch regulation. Cell 173 (6), 1370–1384.
Tabansky, I., et al., 2015. Implications of epigenetic variability within a cell population for "cell type" classification. Front. Behav. Neurosci. 9, 342–359.
Trahan, L.H., et al., 2014. The Flynn effect: a meta-analysis. Psychol. Bull. 140 (5), 1332–1360.
Wrangham, R.W., 2004. Catching Fire; How Cooking Made us Human. Basic Books, New York.
Wrangham, R.W., Peterson, D., 1996. Demonic Males. Houghton Mifflin, Boston, MA.

Further reading

Camp, J.G., et al., 2015. Human cerebral organoids recapitulate gene expression programs of fetal neocortex development. Proc. Natl. Acad. Sci. U. S. A. 112 (51), 15672–15677.
Lawler, A., 2018. Were nomads the world's first traders? Science 358, 1518–1521.

CHAPTER 2

Babies' development

I have offered anthropological explanations of human social instincts which, I argue, are entwined with the growth of sociability during human development. I now want to discuss this second, developmental element in the skein of explanations that account for human attachment.

As I have emphasized from the start, no one of these explanations is sufficient on its own, and only by viewing them together—as fully coordinate with each other—can we configure the multi-dimensional explanation for human sociality. Viewed from this perspective, the phenomenon extends over vast stretches of evolutionary time and is apparent as well (at the other end of the scale), in how small babies instinctively grab for their mothers.

It may appear that babies grab for their mothers simply because she is there—maybe they would grab for any object within reach. But this is not the case. Babies' movement toward their mother is a social act, one of the first that any human performs. It is an induction into sociality, setting the stage for a lifetime of instant, automatic movement toward others. The mother-baby duality is the first real "group" to which the baby belongs, and its reaching out to the other member of this group is a primitive survival strategy. As with the larger communities that I have discussed, *this* group also provides a means of surviving in the world, and the baby knows this instinctively.

Babies' development is a complicated subject which, fortunately, has been studied intensively. The field is somewhat settled, at least in broad outline, albeit interesting new studies continue to appear. For the purpose of this book, I have not had to range around the breakthrough outputs of cutting-edge labs in order to put together a discussion of current thinking. I have learned a lot by reading. For this chapter, I follow an authoritative text in developmental psychology (Keil, 2014), as well as other related discussions, which set out much of what we now know concerning babies' development of social impulses. These treatments show how rapidly and naturally the infant's social behaviors emerge, based at first on the infant's instinctive—and altogether practical—need for attachment.

Origins of Human Socialization
https://doi.org/10.1016/B978-0-323-85861-8.00012-8

Much of what science knows about babies' responsiveness to other humans is, already, embedded in the knowledge base of attentive parents. What's different is that Keil and others systematize and analyze this knowledge, rather than hurriedly and automatically applying it to demanding infants. They stand back from this knowledge and, in doing so, make abundantly clear that babies are capable of sociability (calibrated to babies' motor capabilities) virtually from birth. Also unlike parents, Keil et al. do not attribute to babies the emotions that parents tend to associate with babies sociability. Parents want to think that babies "love" them immediately, just as the parent loves the child. But Keil et al. understand the need to approach babies' actions as motor-reflex, neurophysiological phenomena.

When, I visited Neonatal Intensive Care Units (NICUs), where prematurely born infants in the earliest stages of development are being maintained—e.g., after only 30 weeks of pregnancy—nothing seemed to be occurring in terms of an infant's social behavior. The babies were perfectly well treated—warmed, fed, and followed medically. But they basically looked unconscious, certainly unresponsive when cared for by skilled and devoted nurses. When I examined them closely, time and again, nothing like a spark of sociality was perceptible. So, the question that I had was *when* do social impulses start and how are they made manifest by a being that is so small, underdeveloped, and barely aware of its circumstances? When does a baby become aware of human beings outside itself that it can reach for, that it needs at some instinctive level? Were the NICU babies still too immature to be sociable, while full-term babies could have exhibited that capacity? I recalled my own three children who, at full term, did show an active interest in others.

Mothers will tell you that the first time they hold their baby, the baby is already reaching beyond itself in some uncoordinated but determined way (Fig. 1). Mothers respond instinctively to a baby's need. The baby is presumably not conscious of the fact that it is displaying a need, but that is my point: its social impulses are instinctive. Such instincts instrumentalize the baby's earliest capabilities for survival. The baby cannot care for itself; without these capabilities it would die; so, *babies have evolved to display social impulses virtually as soon as they are born.*

They do this not just by crying (although crying also expresses need) but by physically attaching to the mother. The attachment is physical, not provably emotional. It relies on the baby's still-developing motor capabilities.

Keil (2014) examines babies' social behaviors by starting with motor development. After all, the earliest social responses of babies are reflexes: the

Fig. 1 An early sign of social development is the baby's preferential gaze toward mom. Measurement of infants' direction-of-gaze comprises a tool in experimental developmental psychological research.

rooting reflex "in which brushing an object against the face causes a newborn to move his mouth toward the object and attempt sucking, and the grasp reflex" (p. 117), in which touching the newborn's palm causes him to grasp tightly, and then the suckling itself. A reflex does not require a conscious action, but it needs an "object" toward which a person—in this case, a baby—can act. The baby's motor reflexes are social by definition. Such reflexes, i.e., the baby's sucking/grasping reactions, define the perimeters of its early social world, which consists of the baby and its mother.

The nature of the behaviors attainable by the newborn depends on (and follows) the changing proportions of the baby's body: prenatally, the head so large in proportion to the rest of the body, changing to increased muscularity as the baby matures from birth to 2 years. Further, the baby's ability to make well-coordinated responses to objects in its environment, including people, depends on the baby's active experience in a normal visual environment. Lab experiments at MIT with kittens by Richard Held and Alan Hein (p. 126), and by Keil at Yale with human infants, proved that "active grasping and reaching" when the infants could see and touch normally are crucial for the development of the motor skills intrinsic to social behavior. (Below, I will discuss ranges within/without normal designations).

Of course, one of the exquisitely social aspects of skill-learning involves imitation (p. 131). Even newborns can imitate certain facial expressions like sticking out the tongue or widely opening the mouth. Andrew Meltzoff and his colleagues at the University of Washington argued that the abilities to perceive another human's action and to produce that action might both depend on similar, overlapping neural mechanisms. Further, there may be developmental social mechanisms similar to the phenomenon of the imprinting demonstrated by a wide range of animals: the "powerful and seemingly irreversible behavioral function which seems to ensure that young animals will stay close to their mothers" (p. 135).

Notice that in these very early instances I have just described, the word 'social' should not be confused with the concept of emotion. In babies' early development, social/motor reflexes and social/imitation function are not based on love, i.e., the sort of intense feeling that parents experience almost instantly and imagine to be reciprocal. Babies are simply doing what babies do, based on how they are wired to respond to keep themselves alive and to further their progressive integration into their immediate environments.

I should reiterate that this book concerns the initial movement of one human being toward another; in the case of babies, we are concerned with the initial movement of someone who has only an initial acquaintance with the world (and with other people in general). Yes, babies' early movements are rudimentary attempts at sociality; they are developmental in the sense that the baby has not yet fully developed; they are mechanical and should not be considered comparable even to the instinctive movements of one adult toward another. Certainly, they cannot be proved to be evidence of love or affection. We cannot attribute to babies—in their very early stages of development—feelings that they may not have.

Indeed, the baby's mechanical responses to its environment make us realize how pervasive the social impulse is. Correspondingly, even as we recognize that "social" may have various lay meanings, they may not always apply to baby behavior when the concept is approached scientifically.

This is not to say, of course, that very young infants do not form bonds. They do. Keil turns to social behavior per se when he considers developing bonds between infants and caregivers (p. 189). Before addressing the psychoanalytic approach of John Bowlby (see below), Keil cites evidence that "infants show a preference for looking at faces over other stimuli and attend to human voices," even before they show clear bonds to specific people. Over the first 6 months they learn to "form bonds to specific individuals and smile most in their presence." Then they will show separation distress when "the individuals with whom they have formed specific bonds leave." The nature of these bonds, however, is fascinating. They are not based on the mutuality that we associate with love or affection; rather, they demonstrate the need for self-protection, i.e., the instinct for survival.

The British psychoanalyst John Bowlby (1969) broke out of psychoanalyst Sigmund Freud's theoretical fixation on *psychosexual* development to offer a factual summary of babies' capacities for social behavior. Bowlby came to be viewed as the father of "attachment theory." When Bowlby theorizes (p. 223) the functions of attachment behavior, he refers to Freud's somewhat obscure and complex sexual paradigms but states that "even in mammals, food plays only a marginal part in the development and maintenance of attachment behavior." Instead, he claims that "the function of attachment behavior is protection from predators." In support, Bowlby cites (i) observations of a large number of animals behaving in the wild; (ii) the fact that attachment can easily be elicited in animals particularly susceptible to predators; and (iii) attachment is elicited at high intensity in situations of alarm" (p. 226).

For Bowlby (p. 265), "attachment behaviours begin as series of reflexes – an ethologist would call them 'fixed action patterns' – neonatal crying, suckling, clinging and orientation to the mother." Bowlby organizes his view of the development of attachment behaviors by identifying four "phases" (p. 266):

- Orientation with limited perceptual discrimination from birth to about 8 to 12 weeks, orienting toward any people in his vicinity by stopping crying, "but his ability to discriminate one person from another is limited to olfactory and auditory stimuli";
- Orientation in a more marked fashion toward his mother-figure, until about 6 months of age;
- Maintenance of proximity to the mother-figure by signaling and by locomotion. Now the infant's "repertoire of responses extends to include following a departing mother, greeting her on her return and using her as a base from which to explore." Bowlby refers to a period from about 6 to 12 months and continuing through the second year;

- Formation of a goal-directed partnership. "Here the child comes to infer something of his mother's set goals, and something of the plans she is adopting to achieve them (p. 267)." The groundwork is laid for a partnership.

As a side point, I note that Bowlby (p. 198) sees the strengthening of all these behavioral transitions as reflecting a continuum from non-human primates. This is consistent with my own argument, which encompasses the evolution of *Homo sapiens* as well as that of other species. I will make the point again and again in a variety of contexts that evolution is parsimonious, at least to the extent that it does not "reinvent" successful brain mechanisms and adaptation strategies. This is not to say that everything is the same across species but, rather, that there is still a lot basic conformity, i.e., evolution is "conservative" to the extent possible.

By about 5–6 months, the infant begins to show a strong specific *attachment* to a primary caregiver. The infant starts to differentially direct certain behaviors, including orienting toward and signaling to specific individuals in its environment. Such behaviors serve a protective function for the infant by promoting their caregivers' nurturing behavior. By the end of its first year, the infant demonstrates preference for their caregiver over others, seeking contact especially in times of stress, and reacting with distress upon separation.

Further, Bowlby (p. 199) states that "by about six months baby is starting to respond differently to mother than to other people. When he sees her he will smile, vocalize, and follow her with his eyes; all of these more than with other people. Also, if he is crying he may continue to cry if held by someone other than his mother and stops crying when taken up by his mother. Then he may start crying when his mother leaves the room, but not when others leave. When he begins to crawl, he preferentially crawls toward his mother. By 5 or 6 months he may climb over his mother and explore her face, hair, clothes, but not so with other people. He may use his mother as a base from which to explore. His mother is a haven of safety." Of course, the point is that according to Bowlby, attachment at this stage is a type of emerging preference associated with the baby's need for (i) protection and (ii) accommodation to its immediate environment. Both are necessary for the baby's survival. In this posture, the social impulse—still highly mechanical, albeit coupled with a type of specific recognition of another human being—is a feature of the baby's developing brain mechanisms that will help it survive. Sociality and survival are two sides of the same coin.

Bowlby has been quoted as saying "All of us, from cradle to grave, are happiest when life is organized as a series of excursions, long or short, from the secure base provided by our attachment figures." Certainly, that statement holds for the infants that he studied. But it is important to conceptualize what such attachment means in context with any analysis of the human social impulse. Bowlby, as well as the others cited here, demonstrate that from the earliest moments after a baby's birth it is already *moving toward* another human, however rudimentary and non-emotional that movement may be.

The implications for my three-pronged theory of human sociality are therefore clear: the theory is three-pronged ab initio. Babies do not start becoming social somewhere down the line as they "get to know" other people. They express themselves socially right away, in ways that adults can identify with because we are talking about the first instant of sociality, i.e., immediate reactions that do not require any emotional motivation. We are talking about initial, other-directed movement per se, nothing complicated—the way an adult would instinctively move toward another that they did not know but who needed a hand. The difference is that babies' movements are focused on self-preservation. They are inquisitive, just because everything is new. But at the level that concerns me in this book, they are social actors from the get-go, i.e., quintessentially part of the three-pronged model on offer right from the start.

Babies are thus a curious mix of near-total narcissism—unable to imagine a world outside their own bodies, as far as we know,—and extreme sociality, aware at some instinctive level that but for their mothers they could not survive. The two conditions are inextricably related: it is *because* the baby's awareness is so limited that it must be a *natural* reach for its mother. Only as the baby's awareness expands will it reach for other objects, although its mother still represents the other half of its group and, hence, its means of support. Babies tend to play by themselves for a long time after they are born, and only later come to understand that there are other children with whom they might form a group (however short-lived and provisional). The baby's self-absorption draws it toward the mother—a paradox of nature that also makes perfect sense.

Born social?

In *The Interpersonal World of the Infant*, Daniel Stern et al. (1985) states that "even infants younger than two months tend to shift their gaze from the periphery to the internal features of the face." (p. 52). By 4.5 months, the baby is "coming to a sense that he is a separate physical being from his mother" (p. 53).

Other research supports this idea. Farroni et al. (2002) at the University of London studied infants that were so young—2 to 5 days after birth—that it is difficult to imagine how they conducted their work. However, they observed that "from birth, human infants prefer to look at faces that engage them in mutual gaze and, from an early age, healthy babies show enhanced neural processing of direct gaze. The exceptionally early sensitivity to mutual gaze demonstrated in these studies is arguably the major foundation for the later development of social skills." These same scholars (Farroni et al., 2005) reported that babies from a half day of age up to about 6 days would show *about 2X the preference for an upright face-like drawing compared to upside-down.* "Most researchers agree that in their natural environment human newborns preferentially orient toward faces." Valenza et al. (1996) agreed. The newborns in their study showed "a visual preference for the stimulus for which components were located in the correct arrangement for a human face," and directed their attention away from non-facelike stimuli.

Likewise, babies display a preference for images of biologically meaningful motion of the sort that people actually undertake. Two-day-old babies were able to discriminate among motions that mimicked normal biological motion when first exposed to them, and they "selectively preferred to look at the biological motion display" (Simion et al., 2008). Further, "this preference was also orientation-dependent: newborns looked longer at upright displays than upside-down displays." Additionally, newborn infants "will follow a slowly moving schematic face stimulus with their head and eyes further than they will follow scrambled faces or blank stimuli" (Johnson et al., 1991). In other words, babies are drawn to movements and formations that look like they are associated with humans.

How should we further conceptualize these preferences of babies? At the recommendation of Harvard professor Elizabeth Spelke, I read an important paper by Susan Johnson et al. (2010). At a high conceptual level, Johnson inferred unique patterns of expectations by babies: (i) "Securely attached infants expected infants to seek comfort from caregivers and expected caregivers to provide comfort. (ii) Insecure-resistant infants not only expected infants to seek comfort from caregivers but also expected caregivers to withhold comfort." Johnson and colleagues reached the conclusion that infants form internal working models of attachment that are expressed in the infants' behavior. That conclusion demands a high level of cognitive "buy-in," mental investment, by very small babies, so I am not sure whether I can agree.

Michael Tomasello takes an evolutionary approach to the human (and. Human infant) social instincts that I am examining. Tomasello created an outstanding program of research on infants while leading a Max Planck Institute in Leipzig Germany and now is a professor at Duke. He writes that "Humans' last common ancestors (like chimps and bonobos) had long lasting social relationships" like kinship (Tomasello, 2019, p. 13). Further, he speaks of "uniquely human adaptations for cooperation and shared intentionality" when considering young children's behavior (p. 228), and an "intrinsic motivation to help and cooperate" conferring an evolutionary advantage. I appreciate this position, and have argued (Pfaff, 2015) that human altruistic behaviors can follow naturally from well-understood brain mechanisms.

Tomasello follows the development of infants' social behaviors in a manner consonant with my preceding discussion. From around 9 months of age, they display "shared intentionality" (p. 46) and joint attention (p. 56), i.e., when "infants understand others as intentional agents who have goals and perceptions toward the world." He supports the findings of Colin Trevarthen (p. 54), when infants do "turn taking in emotional expressions toward each other," and (Stern et al., 1985) when they exhibit "emotional attunement in which infant and mother mirror each other's emotional intensity and valence," smiling and laughing. I was struck by his report (p. 224) that at 12 months infants may "approach and console other crying infants." Later, an "infant can establish common ground with [an] adult" (p. 60), when an "adult stopped and pointed toward a target toy, which the infant then tidied up into the basket." Babies point: at around 9 months to request help (p. 98), then as an imperative and to express their attitudes.

Pointing is, in fact, an interesting social behavior. Babies point as a way of sharing: they call attention to something that has caught their eye, and want everyone around to share what they see. There is a famous picture of the British royal family on the balcony at Buckingham Palace, during which infant Prince Louis is pointing to the jets that are streaking overhead. His pointing is a quintessentially social act. This very young child is sharing his delight and curiosity. On a humbler level, we see such behavior all the time when babies, perched on top of supermarket carts, point at items on the shelves while their mothers nod. The babies are not asking for whatever colorful objects they see—they probably do not even know what they are—but, rather, they are calling attention to what they see so that their mother can see it too. Pointing is associated with babies' sense of discovery.

In a sense, this pointing by babies is on a very long continuum with why I set about writing this book and then sought to share it with readers. Writing a book, though it entails thousands of hours of individual research, is ultimately a social act that is a much later (and hopefully more sophisticated) iteration of a baby's pointing behavior. The baby develops its sociality through curiosity as it begins to recognize that a world exists beyond itself. Its social impulse expands beyond reaching for its mother and, as this happens, the baby shares its discoveries as a way of creating a wider social group. As the child develops language, the pointing is replaced by verbal notification … even, perhaps, by (someday) writing books.

Tomasello's book (2019) demonstrates the wide agreement among researchers over time as they have charted the beginning of infants' social lives. My point in citing these naturalistic, experimental studies has been to tell—and reinforce—the story of how rapidly and naturally babies' social capacities develop after birth. But some of those capacities are present already at birth. They are evident immediately. Babies do not provably know that they are being social, and will slowly learn the meaning of the concept. But they do know how to *behave* socially. The difference simply says that such behavior is instinctual.

What fascinates me about the theory that I am offering is that so much of human sociality is beyond our control—certainly in its initial phases. *Even as adults, we don't know that in the instant that we act, we are being social.* We just act. We do not stop to analyze. Babies exhibit this pattern because it is the basic human pattern.

Individual differences

Across all recent research, two major facts stand out. First, babies' earliest social instincts depend on biological needs, which mothers can satisfy. I discussed the mother's role in that regard above. Second, among human babies there is tremendous variability in the rate of social development. The findings of Jerome Kagan, professor of psychology at Harvard, highlight this variability. Kagan (2013, p. 27) states that "all beginnings are interesting, but the origins of talents and traits of children…" elicit a special curiosity. Indeed. That initial "beginning" comprises a major element of this book. I want to know how we got started being social. Moreover, my address to autism involves children's initial development and the congenital failure of their abilities to connect.

When Kagan and his group focused on variations among infants in their emotional behavior, they studied "healthy, white middle class infants born to caring mothers" (p. 57). Here are the extremes. "About 20 percent of the infants were highly reactive. They showed "vigorous movements of the arms and legs, arching of the back, frequent fretting and crying in reaction to a set of unfamiliar events." At the other extreme were 40% who "lay still and rarely cried or arched their back." These were called low-reactive.

Kagan's point is that these same children studied at age 18 had retained these divergent characteristics. As adolescents, high-reactive kids "worried excessively about improbable events." By contrast, low-reactive kids "became relaxed, fearless toddlers and minimally anxious adolescents at eighteen years of age." Thus, the social impulse is unevenly spread across a range of children. The low-reactive kids still survived because they have developed survival-related skills; no doubt, their parents paid close attention to their progress, especially while they were very young. In part, Kagan et al. (1980) appears to subsume the collective findings of developmental psychologists through the time of his research, and none of it undermines current theories about babies' development of increasing sociality. He just demonstrates that this development occurs at different rates which continue to bear on a child's later development.

Emphasizing the continuity of the infant's developmental processes, he seeks to show "how the experiences of the infant exert a profound influence on its development" (p. 2). Further, he assumes that "experiences during the first year extend far into the future: and alter subsequent processes that will occur years later." Third, he considers the mother-infant relation and understands that most scholars believe that "nature intended each infant to be close to its biological mother." He takes Bowlby's concept of "attachment" to mean "the emotional bond between infant and mother." All of this agrees with previous developmental psychologists that (p. 5) "development connotes orderly, organized change toward stronger, more efficient bonds with the mother."

BUT Kagan's real interest is in variations among individual babies (p. 16) and, in particular, the damaging effects of negative experiences during infancy on later psychological processes. Some adults were unloved as children. Some had been subjected to severe stress as infants. Kagan wants to know the consequences (p. 133). What about those "who had a dearth of normal experiences without which the development of human qualities is impossible?" (p. 143). After all, University of Wisconsin professor Harry Harlow demonstrated that baby rhesus monkeys deprived of their mother developed as depressive and asocial. Children who had suffered lonely upbringings in Romanian orphanages had significantly increased rates of autism. Kagan argues that babies can recover: "Children who had been adopted from a Lebanese institution for foundlings before their second birthday (with developmental quotients near 50) and placed in good foster homes, eventually attained average IQ scores of approximately 100, in contrast to those who remained in the institution whose test scores continued to reflect profound mental retardation" (p. 143).

So, these studies all showed data which suggest long-term behavioral implications of difficult early childhoods. But that is not always the case.

On a confusing note, I was surprised to read that some of Kagan's own work (p. 147) failed to reveal a continuity of temperamental qualities between infancy and older ages. For example, he studied 10-year-olds in the Boston area whose "attentiveness, vocal excitability, activity level, irritability and disposition to smile" had been recorded at 4, 8, 13, and 27 months of age. These measures taken in infancy (less than 1 year old) did not predict their dispositions at 27 months. I simply do not understand how this result comports with Kagan's later experiments as discussed below.

Developmental psychologists show enduring respect for a pioneer, Mary Ainsworth, who invented the "Strange Situation" procedure, an experimental method to test the strength and styles of infants' attachments to their primary caregiver (Ainsworth and Bell, 1970).Later, (Tracy and Ainsworth, 1981) she developed a system of classification of babies' behaviors which emphasized a pattern—"anxiously attached and avoidant"—where mothers performed hugging/cuddling proportionally less. This pattern might correspond to Jerome Kagan's "inhibited" youngsters.

Fox et al. (2015) wanted to take measurements of infant temperamental biases assessed at 4 months and compare them with estimates of inhibited behavior during the first 2 years of life. They discerned a self-consistent pattern of results: infants of both genders who displayed high levels of motor activity and distress regarding unfamiliar events were more inhibited at 14 months of age.

How much do these temperamental differences last into early adulthood? In particular, are there neurophysiological correlates? Schwartz et al. (2003), working with Kagan, knew that "Infants with an inhibited temperament tend to develop into children who avoid people, objects, and situations that are novel or unfamiliar, whereas uninhibited children spontaneously approach novel persons, objects, and situations." These authors and others hypothesized that lasting differences in amygdala neurophysiology, especially amygdala neuronal responses to novelty. Schwartz et al. reported that "adults who had been categorized in the second year of life as inhibited, compared with those previously categorized as uninhibited, showed greater Functional Magnetic Resonance Imaging (fMRI) signal responses within the amygdala to novel versus familiar faces." Thus, we can count on the fact that social variations among babies can translate into social variations among adults.

Schwartz et al. (2012), aware that subdivisions of the amygdala play central roles in the registration of fear and subsequent changes in the autonomic nervous system, incorporated amygdala electrical responses into the protocol for studying the implications of infant emotional behavior as studied by Kagan. "Four-month old infants called high reactive (HR) demonstrate a distinctive pattern of vigorous motor activity and crying [in response] to specific unfamiliar visual, auditory and olfactory stimuli in the laboratory. Low-reactive infants show the complementary pattern. Here, we demonstrate that the HR infant phenotype predicts greater amygdalar reactivity to novel faces almost two decades later in adults. Males who were HR infants showed particularly high levels of reactivity to novel faces in the amygdala that distinguished them as adults from all other sex/temperament subgroups, suggesting that their amygdala is particularly prone to engagement by unfamiliar faces." It seems impossible to avoid the speculation that these individual differences in infant behavior, lasting into early adulthood, have implications for some aspects of adult social behavior.

Of course, while my point is that humans are social from "Day One," how they *develop* socially is set in motion (along differing paths, from child to child), launching from the earliest manifestations of the social impulse. It is not enough to say "Oh, brain mechanisms control everything." Following Kagan, brain mechanisms set in motion behaviors that can vary from individual to individual depending on how these individuals' earliest social relationships affect their development. The last thing any neuroscientist would suggest is that social development is predetermined. Humans are not robots. We start with brain mechanisms which are designed to produce prosocial behavior. In many cases, this behavior follows a predictable course. But not always! The fact that we have evolved to behave socially is not less of a fact because some individuals are different from others.

Even better

How well-developed can babies' social behavior be? At a very young age, is it ever more than just reaching and grasping? In my view, no one has made a better attempt to answer that type of question than Karen Wynn, a professor of psychology at Yale. For example, her team (Hamlin et al., 2010) performed two experiments which indicate that "even 3-month-old infants evaluate others based on their social behavior towards third parties." Infants looked at "good puppets" which helped the central puppet in a task significantly more than they looked at "bad puppets" who frustrated the central puppet in the task (Fig. 2). Impressed by the way she had designed her experiments, I visited Yale and witnessed one of these experiments. Every time, the experiment was carried out perfectly and the baby's response was unambiguous: a 3-month old infant essentially could evaluate social behavior and preferred good behavior over bad. Following up that work, Hamlin and Wynn (2011) generated data which replicated the earlier finding and indicated further that the infants' preferences were based on the specific nature of the actions. Thus 3-month old babies preferred the Giver puppet

Fig. 2 Very young children choose the "good" puppet significantly more often than the "bad" puppet in developmental psychologists' experimental puppet scenes.

over the Taker puppet ($p < 0.01$), but there was no significant difference in the experimenters' various conditions. Tomasello comes to a similar conclusion ([Tomasello, 2019], p. 225): "human infants begin actively helping others with their instrumental problems from at least 14th months of age. Even younger than that they discriminate helpers from hinderers in perceptual displays." Wynn puts her findings and those of Warneken and Tomasello in a biological framework: "Evolutionary considerations of the origins of altruism and cooperation emphasize the benefits of reciprocal altruism to all members of a social group, and the benefits to kin of non-reciprocated altruism."

In effect, the Wynn/Tomasello research supports the idea that babies are equipped with social instincts. Babies that they worked with made instinctive judgments with unmistakable prosocial implications, even though no one had taught them—and they had no capacity to learn—the difference between generous and greedy behavior. The babies preferred behavior that was broadly useful in a social context, even though such behavior might be regarded as beneficial to the greedy actor (and therefore have some identifiable utility). I want to stress that these babies were not making moral judgments or *any* judgments. Their reactions were immediate and instinctive, and can be compared to the reflexes and movements described above. They came from brain mechanisms that evolved to produce just this type of instant initial attraction.

Social problems, autism

On the other hand, some developmental psychologists have studied the development of social behaviors in order to try and discern points in the process where things might go wrong.

As much as we applaud the development of prosocial behaviors, e.g., those behaviors studied by Karen Wynn and Michael Tomasello, and those behaviors' adult manifestations, the earliest roots of a social aloofness typical of autism are now a major focus of medical studies. Such social problems do start early, thus, I do not want to wait until Chapter 9 to discuss two aspects of autism that are of immediate importance for my argument: sex differences and the quest for early diagnoses. I have already mentioned sex differences to a degree in Chapter 1, and will get back to them later, but here I approach the issue from a different perspective, not focused on genetics but, instead, emphasizing what can go wrong developmentally.

Early diagnoses. The first time I heard Ami Klin, an Israeli-born psychologist, speak about autism, I knew that I was listening to a master thinker. So, it comes as no surprise that his team at Emory University, led by Warren Jones (Shultz et al., 2018), has clearly summarized not only the normal developments of social behavior, reinforcing the statements at the top of this chapter, but also conceptualizing what goes wrong in autism. They are masters at developing behavioral tests which might provide for early diagnosis.

Klin states that "The infant enters the world in a state of utter fragility, and requires a parent or primary caregiver to provide for survival needs. In playing that role, the caregiver serves as both partner and facilitator, matching his or her own behavior to the needs of the infant in a manner that serves as the foundation for further acquisition of abilities. The behaviors of infant and caregiver are thus mutually adapted and mutually reinforcing." This is interesting, in that it suggests that infants' social reflexes have evolved to promote maximum interaction with—and maximum attraction to—the caregiver that it needs to survive. This type of co-evolution is common across species, such that the colors of certain flowers have evolved to attract insects with a certain range of color vision.

Shulz et al. agree that neonates start with "reflexive and reflex-like patterns of behavior that enable close physical contact as well as distinct patterns of preferential orientation to the sights, sounds, and smells of their caregivers." They state that "already at 10 minutes postpartum, neonates will track a moving face-like pattern farther than a pattern of scrambled or inverted facial features." The baby's preferences for orientation toward eyes, faces, and biological motion have the effect of encouraging social interaction. The baby's crying signals the need for proximity to the mother.

Most interesting, Shulz et al. detail in semi-quantitative terms, the behavioral transitions during the first 6 months of life—from vacant staring to playing peek-a-boo—which comprise typical development of babies' social behaviors. The point is that rudimentary motion toward another human becomes refined over time. Babies become aware of their interaction with others, as in peek-a-boo, and can even control it. This transition from rudimentary to sophisticated is, however, only qualitative. From a quantitative perspective, the instinct is there all along.

So how does this development of the social instinct affect current thinking about autism? Klin et al. (2015) perceptively state that "Efforts to determine and understand the causes of autism are currently hampered by a large disconnect between recent molecular genetics findings that are associated with the condition and the core behavioral symptoms that define the condition." Over the past ten or so years, I have heard so many geneticists try for simple-minded gene/behavior correlations involving autism that I can resonate with Ami Klin's exasperated statement.

Likewise, in a groundbreaking *Nature* paper, Warren Jones and Klin (2013) reported that infants who will later be diagnosed with autism begin with normal eye contact, but "exhibit mean decline in eye fixation from 2 to 6 months of age, a pattern not observed in infants who do not develop ASD." This is the earliest behavioral disability manifest in autistic children.

Following up, Bradshaw et al. (2019) examined early attention in infant siblings of children with autism spectrum disorder (ASD) who are at risk for social and communication delays. "Visual and auditory attention was mapped from age 1 week to 5 months in infants at familial risk for ASD and low-risk typically developing infants. At 12 months, a subset of participants was administered assessments of social communication and nonverbal cognitive skills. A key finding was that high-risk infants exhibited lower performance on attention tasks at age 2 and 3 months compared to low-risk infants." These measurements predicted cognitive developmental differences at 12 months. Early diagnoses using tests like these could permit early therapeutic intervention. That is, since the social instinct is initially present, early detection of its decline could to stabilize the individual and perhaps limit further loss of sociability. No one thinks that any such intervention would be easy, i.e., that there is some straight bright line between detection and cure. But since we can observe early signs of the instinct in an individual, we might begin (perhaps based on those signs and their subsequent weakening) to arrest its decline.

However, we do not yet know enough to connect, comprehensively, a behavioral disability associated with autism to specific genetic and physiological events in the developing brain. In a later chapter, I will consider such an array of genomically regulated developments in the human forebrain, that you will easily see oversimplified "explanations" of autism to be misguided. Nevertheless, very smart neurobiologists still try.

Morten Kringelbach, professor at Oxford, is an expert on the neurobiological connections in the forebrain associated with human emotional (and therefore social) life (Kringelbach, 2016). In Parsons et al. (2010) he attempts to systematize mechanisms underlying six major components of the relationship: (1) orienting system; (2) recognition system; (3) intuitive parenting; (4) attachment relationships; (5) inter-subjectivity; and (6) higher socio-emotional and cognitive functions. Kringelbach cites the relationship that I have been discussing here: "The early parent–infant relationship is defined by the immediate propensity of infants and parents to be attracted to and seek contact with one another. This *orienting system* serves to bring about close proximity between the two members of the dyad." That is, the infant displays an "immediate" capacity for social attraction, which is focused on its parents.

He notes further that "The general orienting response is quickly superseded by a more selective, individuated, *recognition system*. Both parent and infant are selectively responsive to each other's unique characteristics, *either from birth or very shortly thereafter*. Mothers can accurately recognise their own infant early postpartum, even on the basis of single non-visual cues, such as smell, cry or touch. *Within the first few days and weeks of life, infants in turn demonstrate clear preferences for* their mother above other individuals. Infants preferentially orient to view the mother's face rather than the face of an unknown woman, show preference for the mother's voice and even preference for the smell of the mother's breast milk. This recognition system is likely to ensure that the parent and infant seek each other out, thereby maximising contact and facilitating interaction." The child's attraction might be compared to the tropism of flowers that seek the sun and follow it throughout the day because the sun helps them photosynthesize. It is easy and slightly poetic to make such a comparison—such attractions are all over nature and help an array of organisms survive.

By implication, combining Kringelbach's neurologizing with Ami Klin's data, things can start to go wrong quite early in the brain of the baby who will be diagnosed with autism. But while Kringelbach would focus on cortical mechanisms, I am sure that things are more complicated. First, the major and most stable biological fact about autism, the sex difference, will be highlighted in a later chapter and almost certainly starts with sex hormone effects in the brainstem. Second, neuroanatomist Helen Barbas, professor at Boston University School Medicine, and I have found links in subcortical motor control systems which we think are not developing normally in the autistic brain (Pfaff and Barbas, 2019).

If this is a brief introduction to two major features of autism research, the quest for early diagnoses (just mentioned) and now sex differences, Chapter 9 will provide a more inclusive, theoretical treatment of autism.

Sex differences. The most constant biological factor in autism is the huge sex difference in its occurrence: Boys are diagnosed with the condition about four times as frequently as girls. *Here I argue that the robustness of the sex difference, replicated from lab to lab, year to year, and country to country, derives from the multiplicity of hormonal and neurological causes which operate independently and cumulatively.*

During the early days and months in the baby's life considered in this chapter, several factors affecting sex differences in behavior are already operating in the baby's brain. At least nine cumulative aspects of hormonal and brain dynamics guarantee the robustness of the sex difference I autism. First, consider the SRY gene.

- On the Y (male) chromosome, SRY gene expression triggers the cascade of Sox genes. As a result, the medulla of the indifferent gonad develops into testes, leading in turn to testosterone secretion. This, the simplest explanation of the sex difference in autism, led us to the "3-hit" hypothesis, which states that the combination of (i) an autism-related mutation, (ii) prenatal or perinatal stress, and (iii) exposure to testosterone combine to predispose to autism (Pfaff et al., 2011). How does this early exposure to testosterone predispose to autism? Testosterone-fueled arousal signals cause a hypersensitive amygdala which, in turn, leads to anxious avoidance responses to social stress.
  Experimental tests of the 3-hit hypothesis (Schaafsma et al., 2017) made a number of correct predictions. Additional support for the 3-hit hypothesis comes from work with patients (Hanamsagar and Bilbo, 2017; Kopec et al., 2018), rhesus monkeys (Careaga et al., 2017) and laboratory rats ((Scott et al., 2016) and mice (McQuaid et al., 2018). With regard to the stress aspect of the 3-hit theory, Kopec suggests: "Androgen receptors directly bind to the IL-6 promoter and enhance its transcription, and IL-6 can increase androgen receptor expression, suggesting that there may be feedback loops between male sex hormones and IL-6, which could drive the IL-17a response." We note that fetal testosterone (fT) plays a central role in human neurobehavioral sexual differentiation and in later social behavior (Bergman, Glover, Sarkar, Abbott, and O'Connor, 2010; Constantinescu and Hines et al., 2012). Therefore, elevated testosterone exposure during critical periods of early development might be responsible for permanent behavioral changes. These hormonal influences have been demonstrated primarily in individuals with ambiguous genital development (e.g., congenital adrenal hyperplasia). Recently, several studies examined the link between autistic traits and fT levels and suggested that the brain basis of autistic traits may reflect individual differences in prenatal androgens and androgen-related genes (Auyeung, Taylor, Hackett, and Baron-Cohen, 2010; Knickmeyer, Baron-Cohen, Auyeung, and Ashwin, 2008).
- Second, other genes on the Y chromosome (Bellott et al., 2014; Cortez et al., 2014) have also been explored. Some of these genes have important regulatory roles and are expressed in brain. But we don't know how that leads to autism.

- Third, simply the number of X chromosomes: Since males have only one, X-linked genes can be expressed only from X_{male}, whereas females may express X-linked genes from both X's. Of special interest is regarding the role of testosterone is the Androgen Receptor (binds testosterone), an X-linked gene (reviewed Schaafsma and Pfaff, 2014). That could be related to the first causal route mentioned above.
- Fourth, the phenomenon of "X chromosome inactivation." This is complicated. This is tough. In the female, so called "gene dosage compensation" is necessary for the female with two X-chromosomes and can be carried out by inactivation of genes either on X_m or X_p. Such inactivation can be balanced equally between the two X's or can be skewed. Moreover, some genes may escape inactivation. All of this is complex, and was explained by UCLA professor Arthur Arnold (2012). I must say that X chromosome inactivation is so complicated that we don't know how it is involved in behavior.
- Fifth, Parent-of-origin allelic imprinting of specific genes. University of Cambridge professor Barry Keverne (2015) has pointed out important roles for this genomic phenomenon in the production of sex differences in behavior. Here is how it works: Genomic imprinting involves the silencing, or deletion, of alleles, single copies of genes, from one parent or the other. In some cases, scientists might say that the choice of which allele to silence occurs randomly. In other cases, biologists such as Keverne (2015) will argue for the utility of silencing the paternal copy (allele) in some cases and the maternal copy in other cases. As logical and interesting as Keverne's thinking may be, we are still waiting for direct proof of genetic imprinting's effects on the development of autistic behaviors.
- Sixth, the "heterochromatin sink." This means that the inactivated X in the female, usually near the nuclear envelope, can act as a "sponge," thus reducing the concentrations of epigenetically important factors—e.g., histone modifying enzymes—and making them less available to all the genes on chromosomes other than X and Y. Modified gene transcription due to this phenomenon could have powerful effects on behavior.
- Seventh, boys may be more vulnerable to a variety of prenatal insults (e.g., infection, malnutrition, stress) (Gardener, 2011) and, for uncertain reasons, might be more susceptible to genetic mutations relevant to autism (Goin-Kochel, Abbacchi, and Constantino, 2007; Szatmari et al., 2012). Other studies suggest that girls are more resilient to these prenatal, potentially deleterious factors. A popular explanation for such a protective feature is higher exposure to female sex hormones and to

oxytocin at critical periods of fetal development (Carter, 2007; Yamasue, Kuwabara, Kawakubo, and Kasai, 2009).

- Eighth, conversely, the male preponderance in autism could result merely because girls need a larger genetic load to become symptomatic. This is the so-called "Differential genetic load." What that means is that girls, for unknown reasons, may require a larger number of mutations or more powerfully deleterious mutations before they show symptoms. In Chapter 9 I'll discuss a potential female-protective effect.
- Ninth, a special case: oxytocin (OT) is a sexually dimorphic neuropeptide, is known to foster pro-social behaviors, including social recognition, social learning, and reproductive behavior in both humans and animals (Feldman et al., 2013). Furthermore, in rodents, estrogens acting in the hypothalamus and amygdala together with OT reduce social anxiety and foster higher levels of positive social interactions (Lee et al., 2009; Choleris et al., 2003, 2006). A clinical trial done by Parker et al. (2018) found positive results, by demonstrating effectiveness of nasal spray of OT on prosocial behaviors in ASD. Since oxytocin has widely-recognized prosocial effects on behavior, more powerful oxytocin systems in females easily could contribute to the sex difference in autism.

In addition to these biological explanations of the sex difference in autism, consider potential diagnosis biases. In the classroom, for example, the diagnosis bias might occur by a teacher's wanting to get the difficult boy out of the classroom by designating him "autistic" and consigning him to special education, while the (less) difficult girl's behavior is assigned to "individual differences" (Goldman, 2013). Until recently we still labored under Victorian social assumptions that girls are less disruptive, more malleable, and more willing to be coaxed into good behavior. Without recognizing their biases, teachers reinforce these assumptions to the detriment of boys.

Patterns of genes/patterns of behaviors: In no case do neurobiologists who do this type of work claim that any single gene/behavior relationship has exclusive responsibility for regulating the development of social behavior initiation from the baby's brain. Instead, one must talk about *patterns* of gene expression in circuits operating in the vertebrate social brain governing *patterns* of behavior. For example, viewing social behavior as an approach-vs-avoidance decision, Pfaff and Barbas (2019) envision perturbations of dopamine receptor-dependent paths through the basal ganglia as part of the problem in some ASD patients. Different timings of mutations in a *pattern* of cortical neuronal genes and dopamine-influenced genes can help to explain some of the heterogeneity of autistic symptoms.

Stress

There is little doubt that physical and psychological stress to the baby before and after birth have the capacity to interfere with the normal development of prosocial behaviors in the infant. A more recent question concerns how to ameliorate the negative effects of such stress, and some approaches have been developed.

Megan Gunnar, professor of psychology at the University of Minnesota and dean of this field of work (Gunnar, 2017), focuses on the role of social buffering in "reducing the infant's experience of stress and arousal regulation in the context of her early attachment relationships." The success of such buffering bears on the range of individual differences in the ability to handle stress. Such buffering is formalized as in the "Holding Hands" program, which provides innovative preventative intervention aimed at improving the behavioral and emotional functioning of toddlers (McAloon and Lazarou, 2019), using the principles of learning theory. And according to these authors, it works.

Of course, the baby's stress hormones are involved. The early work of Seymour Levine, as reviewed by Suchecki (2018), showed the importance of nutrition in suppressing the corticosterone stress response, and examined how maternal behaviors in laboratory rodents and monkeys regulate specific aspects of the hypothalamic-pituitary-adrenal axis. In contrast, maternal deprivation increased basal and stress-induced activity of the hypothalamic-pituitary-adrenal axis.

In view of our theory that the neuropeptide oxytocin buffers social behaviors against harmful effects of stress (McCarthy et al., 1991), genetic variations in oxytocin signaling systems can play a role (Smarius et al., 2019). Oxytocin receptor gene polymorphisms and oxytocin gene polymorphisms rs2740210 and rs4813627 affected children's anxiety and emotional symptoms upon exposure to verbally aggressive maternal behavior.

One can only hope that these social buffering approaches to reduce the negative impact of neonatal stress will become still more behaviorally sophisticated and will make use of new technology. As I indicate throughout this book, human behavior is continuing to evolve in connection with technology. This does not just apply to adults with i-phones and subscriptions to Skype. Even at extremely young ages, technology is coming to be regarded as an adjunct to behavior. On the individual level, if not always *en masse*, children's behavior can be altered by means of technology. In a way, technology is becoming the drug of choice, replacing conventional drugs that have been used to treat various conditions including attention deficit and autism. The idea that we have "stopped evolving" is wrong. If anything, evolution may be speeded up, albeit by taking new turns. Thus, if the evolution of the human brain is mainly a subject for evolutionary anthropologists, its evolution as the human component of the brain-technology complex is happening before our eyes. The use of simple technologies like iPads with autistic children may provide a good example.

Brain mechanisms

Because of the explosion of interest in the molecular biology of brain development, it has been tempting for some to try for simple genetic approaches to the behavioral developments we have covered in this chapter. For some, a geneticist's thinking goes like this: infants' behavior is changing, the brain is changing at the same time, and the brain regulates social behavior, so why not ascribe these behavioral changes to a gene or two? However, succumbing to the "genetics" temptation is a mistake. Genetic transcriptional events orchestrate so many types of development in the infant mammalian brain—at a minimum, neurogenesis, myelination, presynaptic formation, postsynaptic formations, dendritic growth, dendritic pruning, axon outgrowth, axonal growth direction-finding—that knowing *which* individual steps to try and link causally to *which* elements of behavior creates a scientific nightmare without a more sophisticated approach.

Chapter 11 will cover a more sophisticated approach to genetic issues and, beyond that, "epigenetic" issues as far as they can currently take us. For the development of social bonding, these are dynamic issues.

That said, there are two areas of mechanistic studies of the development of brain and behavior which I can solidly cover here. One is chemical, the other is electrical.

Chemical

The chemical approach most developed, so far, rests largely on the robust field of knowledge concerning a neuropeptide of nine amino acids called oxytocin.

Ruth Feldman at Bar-Ilan Medical School in Israel has produced a long train of papers showing oxytocin's importance in social behavior. At Emory University, Larry Young has discovered some of the cognate molecular biological mechanisms, described in Chapter 11. But first, I discuss the behavior that they considered.

Harel and Feldman (2011) were interested not just in the occurrence of mutual gaze between baby and mother, but also in their timing. "During mother-infant face-to-face interactions, the most dynamic interpersonal context, preterm infants and their mothers displayed short frequent episodes of gaze synchrony." She thinks this synchrony is important for the cognitive baby's development (Feldman, 2006), and I suggest that the persistence of mother-directed gaze by the baby indicates its rewarding properties to the baby and the mother—safety to the baby, satisfaction for the mother (Feldman et al., 2011).

A group at Johns Hopkins Medical School (Nardou et al., 2019) examined reward mechanisms during development which might be relevant to the value of social attention synchrony to the baby. Their paper argued that "developmental regulation of oxytocin-mediated synaptic plasticity in the nucleus accumbens" establishes a critical period for social reward learning. Furthermore, we show that a single dose of (+/−)-3, 4-methylendioxymethamphetamine (MDMA) reopens the critical period for social reward learning and leads to increased oxytocin effectiveness. "MDMA-induced reopening of this critical period requires activation of oxytocin receptors in the nucleus accumbens, and is recapitulated by stimulation of oxytocin terminals in the nucleus accumbens." I think that under difficult circumstances for the baby's social learning, oxytocin opens time windows and increases the probability that the baby's socialization will proceed smoothly.

Recognizing the importance of the mother-infant bond, Atzil et al. (2011) focused on maternal care mediated by the neuropeptide oxytocin. They were especially interested in mothers whose attention to their babies was synchronous with the baby's, and measured oxytocin and gave the mothers brain scans. While observing several "infant-related vignettes," only synchronous mothers showed greater activations in a reward-related nerve cell group, the nucleus accumbens. Additionally, functional connectivity analysis revealed that among synchronous mothers, nucleus accumbens activity and right amygdala were functionally correlated. Correlations between oxytocin and accumbens and amygdala activations were found only in the synchronous group. From this, I can (at least provisionally) conclude that without oxytocin, the baby's socialization with mother would be far less than ideal.

Consistent with the results from the Feldman lab, Elena Choleris et al. (2003, 2006) in my lab discovered linked estrogenic and oxytocinergic systems in the hypothalamus and amygdala which foster normal social behavior. Using an assay of social recognition, we found that female mice whose genes for the neuropeptide oxytocin (OT) or the estrogen receptor (ER)-beta or ER-alpha had been selectively knocked out were deficient specifically in social recognition and social anxiety. There was a remarkable parallelism among results from three separate gene knockouts. Our data strongly suggested the involvement in social recognition of the four genes coding for ER-alpha, ER-beta, OT, and the OT receptor. We proposed four-gene micronet, which links hypothalamic and limbic forebrain neurons in the estrogen control over the OT regulation of social recognition. In our model, estrogens act on the OT system at two levels: through ER-beta, they regulate the production of OT in the hypothalamic paraventricular nucleus, and through ER-alpha, they drive the transcription of the OT receptor in the amygdala.

Thus, among many neurochemical systems bearing on social instincts, the nine amino acid peptide oxytocin and its receptor stand out for their consistency of results over decades. And although this chapter has emphasized forebrain mechanisms, you can be sure that other sites of action up and down the neuraxis will prove important (see New Questions).

Electrical

How much of a role could so-called "mirror neurons" and similar neural systems play in the early development of prosocial behaviors? In a 1991 experiment, neurophysiologist Giacomo Rizzolatti and his collaborators at the University of Parma had spent the morning recording electrical activity from a neuron in a monkey's premotor cortex. The neuron fired high levels of action potentials when its forearm was lifted, as when grasping something. Then, it was time for lunch. But the experimental electrode was still recording from that neuron. Much to the scientists' surprise, the neuron started firing during their lunch. Rapidly, they figured out that the monkey could see an experimenter's forearm lifted, grasping a cup of coffee! After many more experiments they could "report here that many of these neurons become active also when the monkey observes specific, meaningful hand movements performed by the experimenters. The effective experimenters' movements include among others placing or retrieving a piece of food from a table, grasping food from another experimenter's hand, and manipulating objects. There is always a clear link between the effective observed movement and that executed by the monkey and, often, only movements of the experimenter identical to those controlled by a given neuron are able to activate it. These findings indicate that premotor neurons can retrieve movements not only on the basis of stimulus characteristics, as previously described, but also on the basis of the meaning of the observed actions." In other words, monkey-see-monkey-do. More particularly, specific neurons in the monkey's brain allowed it to replicate—or mirror in its own actions—what it had observed in other actors.

Of course, these observations astonished knowledgeable neurobiologists interested in the regulation of social behaviors. Rizzolatti et al. (1996) called the specific neurons "mirror neurons." They referred to some analogous physiological observations in the human brain and proposed that such functionality lies "at the basis of the understanding of (others') motor events."

Following up, Gallese et al. (1996) recorded from a large set of premotor cortical neurons in monkeys, some of which were mirror neurons where "all of which became active both when the monkey performed a given action and when it observed a similar action performed by the experimenter. Mirror neurons, in order to be visually triggered, required an interaction between the agent of the action and the object of it. The sight of the agent alone or of the object alone (three-dimensional objects, food) were ineffective." In approximately 30% of mirror neurons the congruence was very strict and the effective observed and executed actions corresponded both in terms of general action (e.g., grasping) and in terms of the way in which that action was executed (e.g., precision grip). "We conclude by proposing that mirror neurons form a system for matching observation and execution of motor actions." Do neurons like these form the basis of a social response to another person's actions? Could they underlie empathic responses? And how early in life do they show these properties?

The ultimate question, of course is: What about humans? With neuroimaging techniques and even single-cell recording on rare occasions (e.g., Mukamel et al., 2010), researchers have found evidence, in human subjects, for the kind of action-perception matching which should reflect the activity of a mirror neuron system (Salo et al., 2019). Cook et al. (2014) argued that human infants receive enough sensorimotor experience to support the associative learning of mirror neurons.

Rizzolatti (2005) cited further evidence that in the human brain, evidence for mirror neurons is indirect, but positive. He speculated about their functional importance. The most obvious would be to support the capacity for imitation and, more exciting, to provide the capacity for understanding another person's intention. Rizzolatti reviews evidence for the former. The latter stimulates thoughts about peoples' capacity for so-called "theory of mind" (see Chapter 8), i.e., knowing someone else's intentions.

Babies. How much of a role could mirror neurons and similar neural systems play in the early development of prosocial behaviors? One potential function of mirror neurons is imitation. Simpson et al. (2014) used newborn rhesus monkeys to study imitation of facial gestures. Simpson wanted to know about newborn monkeys' visual attention and whether such attention may explain why some infants imitate two facial expressions which can be shared between baby and mother: lip-smacking and tongue protrusion gestures. Lip-smacking imitators, but not tongue protrusion imitators, looked significantly more to a human experimenter's face and to a control stimulus compared to non-imitators. From the perspective of this chapter, the most important finding is that the kind of imitation which could be supported by mirror neurons can be detected in extremely young primate infants.

Further, one could ask about the consequences of excellent neonatal imitation (Ferrari et al., 2009). It turns out that "infants who imitate facial gestures display more developed skills in goal-directed movements (reaching-grasping and fine hand motor control) than nonimitators. These differences might reflect, at least in part, the differential maturation of motor chains in the parietal and motor cortices, which partly overlap with those of the mirror neuron system." Whether the capacity for imitation by babies has anything to do with the development of social cognition is still uncertain (see New Questions). So, we have made some chemical and electrical inroads into the mechanisms for the earliest developments in the infant's social behaviors. But the biggest story will be found in the genetics and epigenetics. They are covered below in Chapter 12.

In this chapter, however, I have sought to demonstrate that babies display prosocial impulses virtually from birth, and that these become more sharply defined and further refined as the baby matures. Babies have evolved brain mechanisms that motivate these impulses, which supports my argument that evolution, brain mechanisms, and babies' development are three inter-related components of human sociality. I should add that none of these components is somehow more important or pronounced than the other—in the case of babies' development, their brain mechanisms have evolved to promote their survival as they grow into their environments. Further, if I have made any salient point it is that babies' development is not just one "separate but equal" branch of humans' path toward social behavior. That is, babies' social development should not be regarded as isolated from the other causal routes. Babies' development is, rather, continuous with adult social behavior as regulated by brain mechanisms that evolved over time.

Humans are humans from babies, to adults, to the sort of techno-humans that we are fast becoming. Though babies are usually studied separately, e.g., by developmental psychologists, they are not separate in any sense that is important here. Their social instincts eventually become those of the adults that they grow up to be. This is why I spend so much writing space on babies, including on the early causes of autism. When William Wordsworth coined the phrase "the child is father of the man" (1802), he did not have in mind the latest neuroscience. But his general idea—that we become who we first were as children, only more so—was right. When considered from this perspective, the human social impulse is not just a complex of factors, but a complex of complex factors starting from day-one that continue to exert influence throughout adulthood with more than a little relation to who we were on day-one.

In the next chapter, I will argue that the fundamental brain circuitry underlying social behavior is actually laid down in fish brains and maintained throughout vertebrate evolution. Human babies are carrying on an evolutionary pattern discernible—with some fascinating complications—across the phylum Chordata.

Perspective

Ethologist Irenaus Eibl-Eibesfeldt (1989) reflects that "most mammalian families include only children of a single age class. The female bears a litter and brings up these young together. The littermates play with one another, practicing specific strategies of intraspecific encounters... Specific play signals inhibit play behavior from escalating into a serious fight" (p. 594). After the young become independent, another litter with no bonds to the first.

"With humans, siblings of different ages are joined in the family unit in a way that distinguishes this particular bonding arrangement from others" (p. 595). Yes, there are tensions, "but mothers generally prepare their children for the new arrival.... drawing out the existing disposition for friendly contact in the form of affection." Eibl-Eibesfeldt provides examples of "pronounced sibling rivalry" in isolated indigenous tribes such as the Yanomami and the!Kung Bushmen, but also describes "the important roles which older siblings can play in the socialization of infants" (p. 600).

New questions

1. The quest for early diagnoses of autism: How early can help be delivered?
2. How will new technologies enable us to better track the baby's earliest social responses?

3. How do mirror neurons in the human brain play into social behaviors?
4. Up and down the neuraxis, from the spinal cord to the forebrain, exactly how does oxytocin facilitate human sociality?
5. Taking off from the suggestions of Harvard professor Elizabeth Spelke, how will psychological studies of babies' sociality be melded with the much larger scientific field of cognitive development?
6. When will we see direct evidence of parent-of-origin allelic imprinting actually affecting the development of social behaviors?
7. What will it mean for childrens' social development to be raised in the time of coronaviruses?

References

Ainsworth, M.D., Bell, S.M., 1970. Attachment, exploration, and separation: illustrated by the behavior of one-year-olds in a strange situation. Child Dev. 41 (1), 49–67.

Atzil, S., et al., 2011. Specifying the neurobiological basis of human attachment: brain, hormones, and behavior in synchronous and intrusive mothers. Neuropsychopharmacology 36 (13), 2603–2615.

Bimbi, M., et al., 2018. Simultaneous scalp recorded EEG and local field potentials from monkey ventral premotor cortex during action observation and execution reveals the contribution of mirror and motor neurons to the mu-rhythm. NeuroImage 175, 22–31.

Bowlby, J., 1969. Attachment, second ed. Basic Books, New York.

Bradshaw, J., et al., 2019. Development of attention from birth to 5 months in infants at risk for autism spectrum disorder. Dev. Psychopathol. 42, 1–11.

Choleris, E., et al., 2003. An estrogen-dependent four-gene micronet regulating social recognition: a study with oxytocin and estrogen receptor-alpha and -beta knockout mice. Proc. Natl. Acad. Sci. U. S. A. 100 (10), 6192–6197.

Choleris, E., et al., 2006. Involvement of estrogen receptor alpha, beta and oxytocin in social discrimination: a detailed behavioral analysis with knockout female mice. Genes Brain Behav. 5 (7), 528–539.

Cook, R., et al., 2014. Mirror neurons: from origin to function. Behav. Brain Sci. 37 (2), 177–192.

di Pellegrino, G., et al., 1992. Understanding motor events: a neurophysiological study. Exp. Brain Res. 91 (1), 176–180.

Eibl-Eibesfeldt, I., 1989. Human Ethology. de Gruyter, New York.

Farroni, T., et al., 2002. Eye contact detection in humans from birth. Proc. Natl. Acad. Sci. U. S. A. 99 (14), 9602–9605.

Farroni, T., et al., 2005. Newborns' preference for face-relevant stimuli: effects of contrast polarity. Proc. Natl. Acad. Sci. U. S. A. 99 (14), 9602–9605.

Feldman, R., 2006. Mother/infant rhythms and behavioral synchrony. Dev. Psychol. 42, 175–188.

Feldman, R., et al., 2011. Maternal and paternal plasma, salivary and urinary oxytocin and parent-infant synchtony. Dev. Sci. 14, 752–761.

Ferrari, P.F., et al., 2009. Interindividual differences in neonatal imitation and the development of action chains in rhesus macaques. Child Dev. 80 (4), 1057–1068.

Fox, N.A., et al., 2015. The relation between reactivity at 4 months and behavioral inhibition in the second year: replication across three independent samples. Infancy 20 (1), 98–114.

Gallese, V., et al., 1996. Action recognition in the premotor cortex. The mirror neuron system and its function in humans. Brain 119 (Pt 2), 593–609.

Goldman, S., 2013. Opinion: sex, gender and the diagnosis of autism—a biosocial view of the male preponderance. Res. Autism Spectr. Disord. 7 (6), 675–679.
Gunnar, M.R., 2017. Social buffering of stress in development: a career perspective. Perspect. Psychol. Sci. 12 (3), 355–373.
Hamlin, J.K., Wynn, K., 2011. Young infants prefer prosocial to antisocial others. Cogn. Dev. 26 (1), 30–39.
Hamlin, J.K., et al., 2010. Three-month-olds show a negativity bias in their social evaluations. Dev. Sci. 13 (6), 923–929.
Hanamsagar, R, Bilbo, S.D, 2017. Environment matters: microglia function and dysfunction in a changing world. Curr. Opin. Neurobiol. 47, 146–155.
Harel, H., Feldman, R., 2011. Gaze behaviors of preterm and full term infants in nonsocial and social texts of increasing dynamics. Infancy 16, 69–90.
Johnson, M.H., et al., 1991. Newborns' preferential tracking of face-like stimuli and its subsequent decline. Cognition 40 (1–2), 1–19.
Johnson, C.S., et al., 2010. At the intersection of social and cognitive development: internal working models of attachment in infancy. Cogn. Sci. 34, 1551–1609.
Jones, W., Klin, A., 2013. Attention to eyes is present but in decline in 2-6-month-old infants later diagnosed with autism. Nature 504 (7480), 427–431.
Kagan, J., 2013. The Human Spark. Basic Books, New York.
Kagan, J., et al., 1980. Infancy. Harvard University Press, Cambridge.
Keil, F., 2014. Developmental Psychology. WW Norton, New York.
Keverne, E.B., 2015. Genomic imprinting, action, and interaction of maternal and fetal genomes. Proc Natl Acad Sci U S A 112 (22), 6834–6840.
Klin, A., et al., 2015. Social visual engagement in infants and toddlers with autism: early developmental transitions and a model of pathogenesis. Neurosci. Biobehav. Rev. 50, 189–203.
Kopec, A.M., Fiorentino, M., Bilbo, S.D., 2018. Gut-immune-brain dysfunction in Autism: Importance of sex. Brain Res 1693, 214–217.
Kringelbach, M., 2016. In: Pfaff, D., Volkow, N. (Eds.), Neuroscience in the 21st Century, second ed. Springer, Heidelberg.
McAloon, J., et al., 2019. Preventative intervention for social, emotional and Behavioural difficulties in toddlers and their families: a pilot study. Int. J. Environ. Res. Public Health 16 (4). pii: E569.
McCarthy, M.M., Chung, S.R., Ogawa, S., Kow, L.M., Pfaff, D.W., 1991. Behavioral effects of oxytocin: is there a unifying principle? In: Jard, S., Jamison, R.L. (Eds.), Proceedings of the Third International Vasopressin Conference. Editons INSERM, John Libbey Eurotext, Paris, pp. 195–212.
Mehr, S.A., et al., 2016. For 5-month-old infants, melodies are social. Psychol. Sci. (4), 486–501.
Nardou, R., et al., 2019. Oxytocin-dependent reopening of a social reward learning critical period with MDMA. Nature 569 (7754), 116–120.
Parker, J.G., Marshall, J.D., Ahanonu, B., Wu, Y.W., Kim, T.H., et al., 2018. Diametric neural ensemble dynamics in Parkinsonian and dyskinetic states. Nature 557, 177–182.
Parsons, C.E., et al., 2010. The functional neuroanatomy of the evolving parent–infant relationship. Prog. Neurobiol. 91, 220–241.
Pfaff, D.W., 2015. The Altruistic Brain. Oxford University Press, New York.
Pfaff, D.W., Barbas, H., 2019. Mechanisms for the approach/avoidance decision applied to autism. Trends Neurosci 42 (7), 448–457.
Pfaff, D.W., Rapin, I., Goldman, S., 2011. Male predominance in autism: neuroendocrine influences on arousal and social anxiety. Autism Res. 4, 163–176.
Rizzolatti, G., et al., 1996. Premotor cortex and the recognition of motor actions. Brain Res. Cogn. Brain Res. 3 (2), 131–141.

Salo, V.C., et al., 2019. The role of the motor system in action understanding and communication: evidence from human infants and non-human primates. Dev. Psychobiol. 61 (3), 390–401.

Schaafsma, S.M., Pfaff, D.W., 2014. Etiologies underlying sex differences in autism spectrum disorders. Front. Neuroendocrinol. 35 (3), 255–271.

Schaafsma, S.M., Gagnidze, K., Reyes, A., Norstedt, N., Mansson, K., et al., 2017. Sex-specific gene-environment interactions underlying ASD-like behaviors. Proc. Natl. Acad. Sci. USA 114, 1383–1388.

Schwartz, C.E., et al., 2003. Inhibited and uninhibited infants "grown up": adult amygdalar response to novelty. Science 300 (5627), 1952–1953.

Schwartz, C.E., et al., 2012. A phenotype of early infancy predicts reactivity of the amygdala in male adults. Mol. Psychiatry 17 (10), 1042–1050.

Scott, A.J., et al., 2016. A neural basis for control of cichlid female reproductive behavior by prostaglandin F2α. Curr. Biol. 26, 943–949.

Shultz, S., et al., 2018. Neonatal transitions in social behavior and their implications for autism. Trends Cogn. Sci. 22 (5), 452–469.

Simion, F., et al., 2008. A predisposition for biological motion in the newborn baby. Proc. Natl. Acad. Sci. U. S. A. 105 (2), 809–813.

Simpson, E.A., et al., 2014. Visual attention during neonatal imitation in newborn macaque monkeys. Dev. Psychobiol. 56 (4), 864–870.

Smarius, L.J.C.A., et al., 2019. Maternal verbal aggression in early infancy and child's internalizing symptoms: interaction by common oxytocin polymorphisms. Eur. Arch. Psychiatry Clin. Neurosci. 121, 43–65.

Spokes, A.C., Spelke, E.S., 2017. The cradle of social knowledge: infants' reasoning about caregiving and affiliation. Cognition 159, 102–116.

Stern, D., et al., 1985. The Interpersonal World of the Infant. Columbia University Press, New York.

Suchecki, D., 2018. Maternal regulation of the infant's hypothalamic-pituitary-adrenal axis stress response: Seymour 'Gig' Levine's legacy to neuroendocrinology. J. Neuroendocrinol. 30 (7), e12610.

Tomasello, M., 2019. Becoming Human. Harvard University Press, Cambridge.

Tracy, R.L., Ainsworth, M.D., 1981. Maternal affectionate behavior and infant-mother attachment patterns. Child Dev. 52 (4), 1341–1343.

Valenza, E., et al., 1996. Face preference at birth. J. Exp. Psychol. Hum Percept Perform. 22 (4), 892–903.

CHAPTER 3

The basic vertebrate social brain—Fish

Fish present complex resource management problems that continue to raise tough biological, economic, and political concerns. The U.N.'s Food and Agriculture Organization has initiated an array of treaties that govern world fisheries. The U.S. participates in bi- and multi-lateral conventions concerning sea life of all sorts—not just fish, but whales and turtles. The issues are invariably sensitive. Where can the British trawl, the French, the Canadians? What can they catch, how much, and when? What do we do about climate change, the destruction of spawning grounds, and the effects of toxic waste? What is the effect of fish farming, building dams, and run-off from power-plants? How do we stop invasive species, introduced when the bilges of cruise ships are emptied? How do we save the dolphins? What type of markets should be permitted in light of health scares from the sale of raw food? We want to maintain stocks but we need to feed people. We think about the balance between finite—even shrinking—aquatic resources and the world's demand for protein.

The issues are so complex and contentious that we rarely stop to think about how, lying just below the surface (so to speak) are questions that broach our understanding of the neural and endocrine phenomena necessary to produce healthy fish. These phenomena affect reproductive behavior—in fish and people—and are thus a component of vertebrate social behavior.

Here, with professor Ishwar Parhar, Monash University, Kuala Lumpur, Malaysia, I talk about fish neurobiology and the basic vertebrate social brain.

The social life of fish—especially as it is abstracted in neural/endocrine phenomena—is key to understanding vertebrate social behavior. Fishes' social life has a direct connection to ours. In terms of circuitry and gene expression, the physiological plan for regulating our social behavior was initially laid down in the fish brain. The *bauplan*, the original structural plan for managing social behaviors among vertebrates, can clearly be observed in the fish brain. In effect, we inherited the fish brain—our brains evolved from theirs. Accordingly, the story of their brain's organization and its effect

Origins of Human Socialization
https://doi.org/10.1016/B978-0-323-85861-8.00011-6

on their social behavior is integral to my three-part discussion of human sociality. It involves evolutionary issues that ultimately connect up with human brain mechanisms and babies' social development.

This surprising claim about the evolutionary history of the social brain actually comprises one small part of a much larger pattern of vertebrate development. For a description of how our bodies' design relies on patterns of genetic expression in fish (our predecessors by about 530 million years) I turned to Neil Shubin's *Your inner fish: A journey into the 3.5 billion year history of the human body* (2009). As the subtitle suggests, our evolution is traceable to much earlier life forms that vastly antedate hominids. By the time that fish (and vertebrates) had begun to emerge, life had already been around for about 3 billion years. After fish/vertebrates emerged, life evolved for about half a billion more years before *Homo sapiens* appeared (circa 200,000 years ago). There was plenty of time leading up to us, for fish to develop the *bauplan* underlying their social behavior.

Shubin reminds us that fish bodies introduced us to left/right symmetry. Fish (and humans) have a front end with sense organs and brains. The spinal cord runs from front to back. The gastrointestinal tract lies inside, a "tube within a tube." Fish (and humans) normally locomote toward the front end. In our early life, as with fish, we develop from three layers: ectoderm, mesoderm endoderm. For fish and humans, the skin and nervous system emerge from the ectoderm; the heart, lungs, and muscles from the mesoderm; our gastrointestinal tracts from the endoderm. All of this is kicked off by temporal patterns of expression of "homeobox" genes which contain the 160-nucleotide base "homeobox" which appear in all vertebrates, different numbers of hox genes in different animals (my lab mice have 39 hox genes).

One can make similar comparisons for sensory organs: *Smell* (Shubin, p. 139) depends on holes in the front end of the body through which we bring air (or water) inside, where the inhaled substance meets a sensitive surface bearing cells which have "olfactory receptors. *Vision* (Shubin, p. 148): light is focused to make the image sharp on the retina (i.e., layers of cells with photoreceptors). Shubin (p. 150) reminds us that this type of visual mechanism is "common to every creature with a skull." Retinal cells express genes for proteins called "opsins," which absorb light and transduce that energy into electrical signals understood by the central nervous system (CNS). *Audition*: Shubin (p. 158) tells us that while fish have no ear flap they do have the bony structure of the middle ear, and observes that (p. 163) "we can trace bones from gill arches to our ears, first during the transition from fish to amphibian, and later during the shift from reptile to mammals." Instead of tiny ears, fish have small organs with ear-like sensory structures "in lines that run the length of the body." The whole system signals vibrations in the environment in a manner analogous to our ears. For the development of this system, the homeobox gene Pax2 is required in both fish and humans.

If we reflect on our structural similarities with fish—not immediately apparent, but obvious when we take a moment—it should not be surprising that this similarity is evident (in a broad sense) in the ways in which our respective brains are designed. Of course, I am not suggesting that fish and human brains are comparable in terms of gross anatomy (e.g., volume), let alone in their rational capabilities (e.g., in a chess match), but I am arguing that they share certain important neural and endocrine phenomena that bear on sociality. You may have heard the expression that an anti-social person is a "cold fish," but this is an insult to fish. As I will demonstrate, fish manifest recognizable social impulses. We rarely observe such impulses (as we might a baby's) because we rarely see fish in the wild. But fish are well-studied creatures whose social life is fascinating. I will examine the basis for their impulse to be social and why that matters to the subject of this book.

Foundations of the vertebrate social brain

About 20 years ago, neurobiologists began to explore the idea that certain groups of nerve cells in the forebrain—and the connections between them—could be involved in social behaviors of all vertebrates. First, Sarah Winans Newman (1999) at the University of Michigan, drew attention to the possibility of a "vertebrate social behavior network." Behavioral endocrinologist James Goodson (2005) at University of California, San Diego agreed. Hans Hofmann, professor of Cellular and Molecular Biology at the University of Texas (O'Connell and Hofmann, 2011, 2012), stated that "behaviorally relevant brain regions are remarkably conserved over 450 million years of evolution." That is, over eons of evolution, certain common structures emerged in the vertebrate brain that promoted social behavior. "From fish to the philosopher," specific types of neurons in these cell groups regulate social behaviors essential for the survival of the respective species.

These scientists were talking about neurons in the ancient telencephalon and the ancient diencephalon. The most famous part of the telencephalon, the highest and most forward part of the brain, is the cerebral cortex, but there is much more to it than that. During vertebrate evolution, as the new cerebral cortex, the "neocortex" got bigger and bigger, older parts of the telencephalon which received important olfactory inputs were forced to roll underneath the cerebral cortex. Importantly, nerve cell groups collectively named "the amygdala" (named for its shape, the Greek word for almond) would survive. Actually, the "amygdala" itself comprises more than 10 cell groups. On the medial side are neurons crucial for sex and on the lateral side are neurons crucial for fear. The medial amygdala is part of the vertebrate social brain network.

An important set of axons exiting the amygdala are called the "stria terminalis." They project medial amygdala signals to the hypothalamus and the preoptic area. Mixed among these axons are cell bodies of neurons which form a so-called "bed nucleus," that is, cell bodies embedded among the axons. This "bed nucleus of the stria terminalis" is also part of the vertebrate social brain.

Farther forward in the telencephalon is the Lateral Septum, so-called because its left and right sides lie close to the "septum" dividing the left and the right sides of the forebrain. The Lateral Septum is still another part of the vertebrate social brain.

Some of the axons running out of the amygdala and bed nucleus of the stria terminalis and the axons running posterior from the Lateral Septum synapse in the Medial Preoptic Area. Others synapse in the Ventromedial Hypothalamus (in the diencephalon which sits between the telencephalon and the midbrain). Neurons in these two groups, the medial preoptic area and the ventromedial hypothalamus, are part of the vertebrate social brain network. Specific types of neurons in these two groups regulate elemental social behaviors crucial for reproduction (see next chapter). In the Anterior Hypothalamic Area, some neurons which express the gene for the neuropeptide oxytocin support friendly social behaviors, while other neurons—a bit more lateral—support aggressive behaviors.

Thus taken together, it clear that: (1) neurons in the medial amygdala, (2) bed nucleus of the stria terminalis, (3) the lateral septum, (4) the medial preoptic area, (5) the ventromedial hypothalamus, and (6) the anterior hypothalamus form the vertebrate social behavior network. Although the teleost homologs of the mammalian amygdala, BNST and lateral septum are still ambiguous, several studies have demonstrated functional conservation of these brain regions using developmental, molecular and behavioral approaches (Wullimann and Rink, 2002; Lal et al., 2018). This is essentially true for all vertebrates, including fish and humans.

These neuronal cell groups, crucial for social behavior, are heavily connected to each other in a manner similar to what was discovered for sex hormone binding cell groups (Pfaff and Keiner, 1973; Cottingham and Pfaff, 1986), as I cover next.

Hormone receptor-expressing cells in neuronal networks

Since we knew that much of the vertebrate body plan has been conserved from fish to humans (Shubin, 2009), we wanted to study sex hormone-binding neurons in fish brains. In the paradise fish brain (Davis et al., 1977), we saw intense binding of radioactive estrogen to receptors in neurons in the medial hypothalamus, even though there was no hormone binding in the midbrain, hindbrain, cerebellum or spinal cord. Also, estrogen receptors were expressed in certain neurons of the preoptic area and in the ventral forebrain in teleosts (Hawkins et al., 2000; Pellegrini et al., 2005). Testosterone binding, revealing androgen receptor expression, was similar to the estrogen results. These results showed expression of the genes for steroid hormone receptors in a limbic/hypothalamic system that seems to be universal among vertebrate central nervous systems (Pfaff, 2017).

What is the meaning of hormone-dependent nerve cell groups projecting to each other (Pfaff and Keiner, 1973)? First, the string of hormone actions in a series of neurons offers the opportunity to *distinguish* the system from other, overlapping systems. If an individual neuron has multiple overlapping responsibilities, its hormone-dependent actions could get mixed up with other actions. But the hormone action applied to the next neuron in the chain and then to the next permits the chain of neurons to stand out in a way that an individual neuron could not. Second, think of the output from a hormone-dependent transcriptional or electrical action in terms of its "signal signal-to-noise ratio." If that hormone signal gets repeated in neurons projecting to the next and still more to the next hormone-dependent neuron after that, then the signal of the final output will be increased while the noise stays the same. Thus, the critical *informational* parameter, the *signal-to-noise ratio*, will be increased. Third, and most exciting, consider that we do not yet know the mathematical functions achieved by these hormone-dependent networks. Are they additive? Multiplicative? Exponential? Suppose for a moment that they are multiplicative. Then if estrogenic or androgenic hormone action doubles the relevant transcriptional or electrical signal output in the first neuron, the second hormone-dependent neuron in the chain has the opportunity to raise the hormone effect from 2× to 4×. And the third hormone-dependent neurons, to 8×. In a word, the hormone-signaling neural net provides *amplification*.

The social behaviors which depend most obviously on steroid hormone receptor expression are those associated with reproduction. Of course, the neuropeptide controlling all aspects of reproduction is the 10 amino acid protein fragment called Gonadotropin Releasing Hormone (GnRH) (Parhar et al., 1995a, 2016). In fish, as in other vertebrates, the neurons expressing GnRH are not born in the brain, as other neurons are, but instead are born in a cell layer the nose called the olfactory placode (Parhar et al., 1995b, 1996; cf. Schwanzel-Fukuda and Pfaff, 1989). From the olfactory placode, GnRH neurons migrate up the septum of the nasal cavity, along the trail of the nervus terminalis. They travel past a neurovascular bundle at the front of the developing brain, migrate along the bottom of the brain and then turn into their functional positions in the preoptic area (Fig. 1). We reason that in the olfactory epithelium, GnRH neurons can sense water temperature, hydrostatic pressure (from the column of water) and pheromones. As a result, GnRH neurons may be directly sensitive to the differences between cold, deep sea water and pheromone-laden river water. Thus, a possible biological rationale for the GnRH neurons' nose-to-brain migration is that it assists the sea-to-river fish migration by some fish species.

Parhar et al. (2016) have described three separate GnRH systems in the fish *tilapia* and their interactions with anther neuropeptide called "kisspeptin." As they note, GnRH effects on social behaviors can be indirect (routed through actions in the testes and ovaries) or directly within the brain. Importantly, in fish, GnRH neurobiology is susceptible to endocrine disrupting chemicals such as triphenyl phosphate (a flame retardant) (Liu et al., 2013, 2016), and this fact suggests that there would be environmental consequences for the animal's social behaviors connected with reproduction.

As a side point, GnRH neuronal migration, the outcome of the 1989 Nature paper referred to above, has been replicated in all vertebrate species tested (Parhar, 2002).

Ultimately, the smooth performance of the social brain in fish as well as in other vertebrates depends on the operation of gene regulatory networks during development (Baran et al., 2017). Charting the gene members of such networks and understanding how they form networks during normal development will occupy molecular neurobiologists for decades. My point here, however, is to stress that there is a remarkable continuity among social brain neurobiology whether the vertebrate be a fish or some other vertebrate.

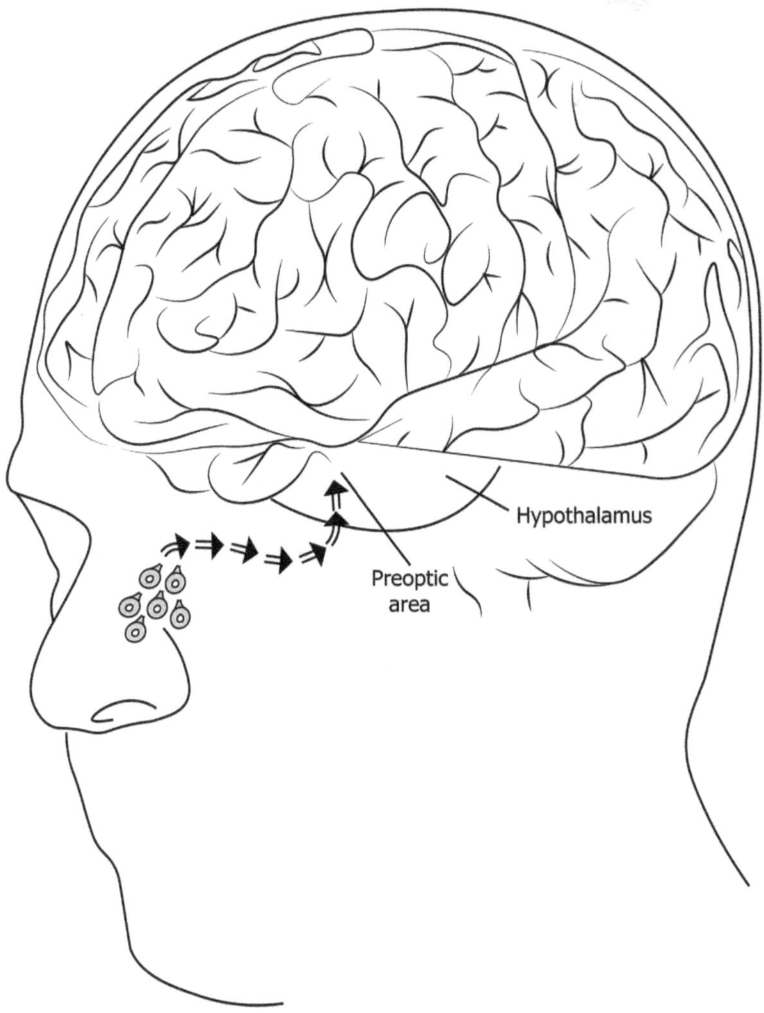

Fig. 1 In all vertebrate species studied, including humans, nerve cells expressing the "master gene" for the brain's control of reproductive biology are actually born in the "nose" (the olfactory placode) and migrate into the brain. Failure of migration can cause loss of libido.

It is perhaps disconcerting to think that, among vertebrates, fish got there first in terms of developing the neurobiology that attracts one individual to another. It is kind of humbling. But evolution *is* humbling. It demonstrates that humans did not just start building pyramids that evolved (over a few thousand years) into skyscrapers. Humans got the benefit of half a billion years of evolution to attain the social brain that is at once incredibly complex and also is just something that we take for granted. The social impulse responsible for the pyramids—and for the preceding pre-dynastic architecture—was around for a very long time already. It is also humbling to think that the variety that we see in nature actually has certain basic, unifying designs that prevent human beings from feeling unique. Humans are highly adapted to their environment because of their social brain, but only because they inherited its proven design from other members of the vertebrate phylum. When it came to the social brain, evolution did not reinvent the wheel when it got around to us.

Effects of hormones and neuropeptides; molecular and behavioral studies

A well-developed field of study has revealed hormonal, pheromonal and neuronal mechanisms for social behaviors in fish (Goncalves et al., 2017). Most of the evidence has to do with elemental social behaviors such as sex, parenting and aggression, but evidence for social bonding has also proved convincing.

Social behaviors in fish depend on the "social behavior network" first proposed, as noted above, by Sarah Newman (1999) at the University of Michigan and extended to fish by John Goodson (2005). This network includes the lateral septum, the medial amygdala, the Bed nucleus of the stria terminalis, the Medial preoptic area, the Anterior Hypothalamus and the Ventromedial nucleus of the hypothalamus. Genes expressed in anterior hypothalamic neurons, arginine vasotocin (homolog of AVP) and gene for isotocin (homolog of oxytocin) play important roles in this network. Communication with neurons involved in behavioral reward, such as in the ventral tegmental area (called the periventricular nucleus of the posterior tuberculum in fish (Rink and Wullimann, 2001; Luo et al., 2008)), play important roles in rewarding social behaviors.

Performance of the social behavior network (SBN) in fish was tracked, e.g., by Teles et al. (2015), by quantifying immediate early gene expression—a sign of neuronal activation. The results showed relatively uniform neuronal activation across the SBN, supporting the notion of its importance in the fish brain.

Regulatory roles for steroid hormones are obvious, as indicated by their binding in the forebrains of fish (mentioned above), and the subsequent transcriptional phenomena are drawn into explanations of the occurrence of mating behaviors. Aggressive behaviors depend on androgenic hormones, but some of their effects are ultimately due to the enzymatic conversion of androgens into estrogens ("aromatization"). For these hormone effects preoptic area neurons are crucial, in part because of the GnRH neurons which regulate the pituitary's release of gonadotropic hormones and in part because of direct behavioral effects.

Beyond these elemental behaviors of great biological importance, social bonding occurs in fish (Godwin and Thompson, 2012). For example, affiliative behaviors and attention to social stimuli seem to be increased by Isotocin (precursor of oxytocin) and reduced by AVT (more associated with aggression). Some fish species display pair bonding, regulated in part by chemical signals such as in urine and featuring both of the release of such a signal in one member of the pair and response by the other member.

Development. A fortunate collaboration between the molecular biological lab of Gil Levkowitz at the Weizmann Institute and Rui Oliveira's teams at three research institutes in Lisbon (Wircer et al., 2017) led to results which show the molecular complexity of neural developmental steps—the types of complex steps which will become ever better recognized in coming years.

Thompson and Walton (2013) had reviewed data on gene expression for oxytocin-like peptides and vasopressin-like peptides in different species of fish, aware of the importance of these peptides for the regulation of social behaviors. A recent study by Mathuru et al. showed that zebrafish exposed to alarm substance preceding being reunited with familiar conspecifics displayed spiked isotocin (IT) levels followed by dampening of circulating cortisol, indicating the involvement of IT in social responses as well as decreasing stress-related responses (Mathuru et al., 2017).

Thinking about the development of socially relevant oxytocin gene expression, Wircer, Levkowitz and Oliveira studied a developmental transcription factor orthopedia (Otp) with respect to hypothalamic development and behavior. In the zebrafish one talks about two forms of this gene, called *otpa* and *otpb*. It turns out that only the *a* form affects stress responses and social preferences. That is, $otpa^{-/-}$ mutants spent significantly less time in the "social zone" of the zebrafish behavioral assay. Mutation of the *b* form had no effect. In this work they discovered a new subpopulation of oxytocin neurons, a group of cells which also express the neuropeptide crucial for stress responses: corticotropin-releasing hormone (CRH). These particular neurons project primarily to the spinal cord and are important for social preference (without affecting responses to stress). Thus, molecular genetic work extended to single neurons and small subpopulations of neurons revealed the unique contribution of a sub-population of hypothalamic neurons in zebrafish.

Phenomena in fish social behaviors

Social effects on reproductive and aggressive behaviors

Among fish, one of the most striking phenomena is that social environment (the presence and behaviors of other fish in a group) can regulate sex roles in individual fish. Depending on the species, females can change into males, and males can change into females (Fernald, 2002; Kobayashi et al., 2013; Liu et al., 2017), i.e., they will display the neural and hormonal changes required for such alterations. Such physiological adjustments involve a wide range of cell groups including neurons, pituitary gland cells and cells in the gonads. Clearly, there is a biological, evolutionary reason for this capability, and it involves maintenance of the optimal sex ratio for the species to reproduce successfully.

Stanford University professor Russell Fernald (2012) analyzed another type of important environmentally-responsive change in a species whose short name is *A. Bertoni*. This was the development of a "non-dominant" into a "dominant" male. Dominant males are reproductively competent. Non-dominant males are not. Here is one of the key social manipulations: In the laboratory, moving a dominant male into a social environment where there is a significantly larger dominant male makes the newly moved male lose its color. When a Dominant male is moved into a social system with larger (> 5% in length) Dominant males, it abruptly loses its color and the new, threatening environment makes it join other non-dominant males and females. Its GnRH neurons in the preoptic area shrink, and so do its testes. The opposite direction of change can also occur. When a non-dominant male is moved to where there is no dominant male and it is the largest, its bright coloration appears within 20 s and then the necessary neural and gonadal growth takes place. Similarly, if a non-dominant male is in a social group where the dominant male disappears, those same social-climbing steps take place. Behavioral steps toward dominance depend in part on female mate choice. "Females choose males displaying the most exaggerated sexual traits—whether behavioral, morphological, or material, such as food and shelter" (Fernald, 2012). Fighting ability also counts. In terms of the social information involved, the striking role-change information involves the fish being able to see each other's behaviors and, in Fernald's terms, make the appropriate inferences. That is, the males are aware of their social environment and change their behaviors to seek gain and avoid loss (reviewed, Maruska and Fernald, 2018).

The *A. burtoni* genome has been sequenced, enabling a class of genetic manipulation experiments not previously possible. (http://www.genome.gov/11007951.) One family of genes we already know to be involved in the non-dominant to dominant reproductive phenotype will be those important for GnRH neuronal development, GnRH1 cell size, GnRH peptide level, and GnRH receptor levels. Recently, Fernald and coworkers have reported gene-knockout of prostaglandin F2 alpha receptor (female-derived sex pheromone) using CRISPR-Cas9 technology and its effect on sexual behavior in this species (Juntti Scott et al., 2016).

My point in describing these physiological transformations among fish is to demonstrate how, at the most basic level, fish display social impulses. Their bodies undergo change on account of—i.e., in response to—the social configuration in which they find themselves. Of course, the fish do not "think" about these changes. They do not consider whether such change is in their best interest. Rather, the changes just happen, triggered by the neurophysiology of the fish's social brain. At this level, it is clear that to be social does not require any awareness of being social. Fish are not consciously social when they change sex or color—they are social just because that is the way their brains are designed to react under appropriate conditions. As I have argued, the human impulse to be social is not, qualitatively, different. Babies' motor response to being touched is unthinking ... but social. To be social, as illustrated by fish and human babies, is to act instinctively through very basic actions in response to external stimuli. In adapting to their environments through a social response, fish are behaving in a way that helps them survive. Babies need to survive, and so they behave in the social ways that babies do, even though they are not provably conscious of why they are behaving socially.

Now I will describe other social behaviors among fish that demonstrate how their social brains operate.

Social decision making

Teles et al. (2015) used the expression of the Immediate Early Gene c-Fos and egr-1 among other genomic analyses to dissect mechanisms for social behavior regulation in zebrafish. Immediate Early Genes such as c-Fos are important because their expression kicks off a large number of other transcriptional events in the cell, and in a wide variety of nerve cells c-Fos expression is correlated with increased electrical activity.

Thus, in order to analyze social decision-making by fish, Teles et al. (2015) created distinct states of social behavior by isolating the fish, having them fight a real opponent (where they can win or lose), or fight their own image on a mirror. Winners only display aggressive behaviors and losers only display submissive behaviors. Mirror-fighters only displayed aggressive behaviors because they fought and could not lose. Expression of *c-fos* revealed distinct co-activation patterns in all social behavior states. That is, each social behavior state has different sets of significant correlations among gene expression levels between different network nodes, indicative of behavior state-specific co-activation patterns.

Social eavesdropping

Fish in groups can gain important information from other fish by attending to those other fish, in effect, by eavesdropping (Abril-de-Abreu et al., 2015). For example, an individual fish can find out about the relative dominance of others in the group and can see how others are responding to changes in the environment. The issue for the human investigator is how to determine what an individual fish is attending to (eavesdropping on). The best clues are how the individual arranges its proximity to the visual targets and the individual's nose-to-tail axis' angle; a fish in a social species is tuned to the exchange of information between conspecifics. These authors found that zebrafish pay more attention to other fish fighting than to other solitary fish, in a manner that would guide the individual's behavior, for example, toward a dominant male. In fact, Lopes et al. (2015) studied the gene expression profile in the brains of zebrafish which differed according to whether they were attending to fish that were or were not fighting. Such transcriptional differences could influence all subsequent social behaviors of these fish, be they aggressive of friendly behaviors. The immediate early genes fos and jun were involved, as well as other genes which could be associated with changes in memory or alertness.

Further, Oliveira et al. (2016) again made clever use of the mirror images of zebrafish whose behaviors were tested and brain mRNA's assayed. In order to track social status-dependent changes in internal state and behavior, they established four behavioral conditions: Isolation, Mirror fighting, Real Fighting/Win and Real Fighting/Lose. More than 100 genes showed different levels of expression in the zebrafish brain, specifically affected by Real Winning versus Real Losing. It was surprising that these significant results did not include any examples in which an individual gene changed in one direction as a result of winning versus the opposite direction as a result of losing. Interestingly, there was a higher impact of losing in comparison with winning on the state of gene expression in the brains of individual fish. The authors suggest that socially isolated fish are a bit more like the winners, thus to explain that the magnitude of the loser effect in behavior is much larger than that of the winner effect.

I should note that when you stop to think about it, ideas of winning and losing are social ideas which, at the most basic level, can operate where survival is in question. Winning and losing are intrinsically social because, as activities, they make no sense except as against an opponent. The fish's social brain is wired for social responses in win-lose situations—A contra B—if not over life and death, then perhaps over a food source. Again, the fish does not consciously think about how to win but, in a social situation against an opponent, it instinctively attempts to win. In this regard, to have a social brain does not always imply that an individual will be kind or "play nice." Rather, it means that the individual will respond appropriately, i.e., as its brain is designed to make it act in a particular social setting. In this sense, the social impulse is neutral. It is about movement precipitated by the immediate environment in which the individual must act in response to another individual.

But in their attempt to survive, fish do more than engage in combat. Like humans, they support each other. In this regard, their social brains promote "social" activities in the positive way that we commonly associate with that term. I describe these activities below, with the caveat that we should not ascribe conscious intentionality to fish.

Social support

What is the mechanism for social support among fish? Faustino et al. (2017) studied a phenomenon in zebrafish that they term social buffering, exposing test fish to an alarming chemical substance which alerted them that other fish nearby had been harmed. In one part of the experiment the test fish were alone; in the other, support fish were nearby or there were olfactory stimuli from other fish. When tested alone, the test fish showed a remarkable degree of fear, displaying a stereotypic response involving erratic movement (as opposed to swimming smoothly in oval patterns), and freezing in place. When tested in the presence of other fish (or other fishes' odors) fear responses were significantly reduced. In fact, individual visual or olfactory stimuli representing other supportive fish were almost as effective as the combination of the two sensory modalities. Thus, a scary situation was socially buffered. The fishes' response was changed because they sensed other fish, i.e., they knew (instinctively) that they were in a socially positive situation. Their social brains interpreted the other fish (or their simulacra) as a line of defense, even though there were no movements signifying a readiness to help. Further, Faustino et al. found that social buffering was correlated with immediate early gene activation of forebrain regions, which they predicted would be true. That is, patterns of gene activation in parts of the vertebrate social brain network discussed above—patterns specific to each experimental treatment—suggested the participation of neuronal groups in this network which could explain the social buffering phenomenon.

Collective cognition in fish?

During TV specials, we have all seen amazing footage in which vast schools of fish—maybe hundreds—suddenly turn on a dime in a totally different direction. It's like every fish in the line, maybe even those far back, simultaneously got a command to change course—and all of them responded with precision. How? Who issued the command, or was it some kind of communal decision in response to a stimulus that all of them processed at once? The ability of large numbers of individual fish to behave in a coordinated fashion, which can feature rapid alterations of behavior in a manner common across the group, has led some theorists to consider the possibility of "collective cognition"—i.e., where each fish somehow knows information at once, and then immediately acts on it. From time to time, neuroscience runs into mystical-sounding ideas like collective cognition which are later subdivided and analyzed to yield testable ideas and proven solutions. The case of the direction-altering fish is one such instance.

Mechanisms underlying "emergent group behaviors" by fish and other kinds of animals (notably birds) have come under systematic study by scientists at the Max Planck Institute in Konstanz, Germany. Alex Jordan, a group leader there, observed that "In hierarchical social systems, such as those found in human and animal societies and digital systems, individuals may differ in their influence on *emergent* group properties (italics mine)." For the neuroscientist, the first question to ask is about the stimuli that one fish (call it a "leader fish")—in response to environmental change—offers to other fish so as to kick off behavioral changes in large numbers of fish. That is, proximate mechanisms lead to behavioral changes widespread across a large group of fish.

In conjunction with scientists in the departments of Physics and Ecology/Evolutionary Biology at Princeton University, Iain Couzin studied fish that swim in large groups ("schools" or "shoals"). Strandburg-Peshkin et al. (2013) considered the importance of the visual system in receiving social signals and compared four theoretical models to conceive the relative importance of such social signals: "Metric: all individuals within a certain distance. Topological: a fixed number of nearest neighbors. Voronoi: those individuals sharing a boundary in a Voronoi tessellation of the group. Visual: all individuals that occupy an angular area on the retina of the focal fish that is greater than a threshold value." The answer was clear. For all models, both the efficiency and the transitivity of social communication simply rose quickly as a function of the number of the focal fish's neighbors.

Sensory stimuli for the causation of rapid changes in groups of fish are not limited to vision. I was alerted to the importance of acoustic communications among fish by reading Scales' (2018) *Eye of the shoal*. Couzin and his colleagues, led by Sara Rosenthal et al. (2015), studied a group evasive maneuver called the "fast-start" behavior, exhibited when the schooling fish perceive an aversive stimulus (e.g., via the visual, acoustic, or mechanosensory systems). In fish, this rapid behavioral response depends on a pair of giant neurons in the hindbrain, the Mauthner cells, which are well known since their likely evolutionary descendants are giant neurons in the (human) medullary reticular formation, which stimulate CNS arousal in mammals (Pfaff, 2006; Calderon et al., 2016; Tabansky et al., 2018).

Thus, the initial stimulus for a sudden change in behavior by an entire school of fish might be visual, acoustic, or mechanosensory. Once initiated by one fish, how might the impetus for behavioral change be transmitted to other fish in the school? It depends on individual differences among fish: "a simple mechanism to explain the emergence of collective behavior *from consistent individual differences*, including variation in the structure, leadership, movement dynamics, and functional capabilities..." (italics mine) (Jolles et al., 2017). That is, given that we know about the sensory cues which govern individual fish behavior during movement of the school, how are signals transmitted from the more sensitive individuals to other fish? Hein et al. (2015) argue that, assuming social attraction between individuals, less sensitive individual fish will nevertheless regulate the distance to the nearest neighbor(s) and, indeed, "climb the density gradient" in the school.

During experiments with 12 species of fish, Hein et al. (2018) used experimental control over an artificial, threatening, "looming" stimulus to measure the dynamics of evasive responses. Evasion responses were "generated through a sequence of well-defined decision rules that convert visual sensory input into behavioral actions." Using an automated system to present visual threat stimuli to fish in situ, Hein showed that "individuals initiate escape maneuvers in response to the perceived size and expansion rate of an oncoming threat using a decision rule that matches dynamics of known loom-sensitive neural circuits." Most important, the "rapid turns moved the animal's head out of the perceived path of the looming object, and are consistent with the prediction that prey should avoid high-speed attacks by turning, a strategy known as the turning gambit." This highly polar turning response resembles responses to looming objects reported in laboratory studies of Mauthner cell- and reticulospinal circuit-mediated escape responses of zebrafish. I believe that the rapid change in behavior illuminated by Hein et al. depends on these Mauthner and other reticulospinal cells and exactly fits the theoretical framework of Pfaff and Banavar (2007) for generalized brain arousal that similarly that supposes a physically defined phase transition in behavior. In our theoretical framework we proposed that CNS arousal systems operate in a critical zone in which a stimulus can trigger a rapid change in behavior.

In terms of how these mechanisms mentioned above might have evolved, Couzin and his Princeton collaborators propose that "natural selection, acting on selfish individuals, results in the highly effective collective responses evident in nature" (Hein et al., 2015). I think they mean that selfish individuals will get left out. These authors write that "that populations evolve toward a distinctive regime in behavioral phenotype space, where small responses of individuals to local environmental cues cause spontaneous changes in the collective state of groups. Again, these changes resemble phase transitions in physical systems," similar to neural phase transitions in CNS arousal mechanisms proposed by Pfaff and Banavar (2007).

Such research is not limited to studies of fish (Nagy et al., 2018). Flack et al. (2018) investigated the group movements of a flock of 27 naturally migrating juvenile white storks. They showed that a small number of leaders navigated to and explored thermals, whereas followers benefited from their movements. That is, the authors examined collective movements in these migratory birds. It turns out that "a small number of leaders explored thermals" on which the birds can rise and save energy, and the followers benefited from their movements. Gautam Reddy Nallamalla, at the University of California, San Diego proposes that vertical wind accelerations and roll-wise torques are effective mechanosensory cues for soaring birds, and provide the basis for a navigational strategy. Then, in collective migration, local-scale leader-follower strategies emerge. This strategy leads to scaling, as discussed below.

In this work and all the research with fish just discussed, it is reasonable to conclude that there is a distribution of sensitivities across an animal population. The animals most sensitive to the physical, environmental variables in question respond individually in an adaptive way. When they do, others—fish or birds—see that they are, and even feel that they are doing so and adjust rapidly. In fact, consistent with behavioral descriptions by signal detection theory, sometimes the highly sensitive individual "responds" (acts) when there is no stimulus at all (a "false positive"). When the fish school (or bird flock) follows, that accounts for the occasions when we see moving populations suddenly change direction for no reason at all.

How does the fish school or bird flock change behavior so rapidly? According to physicist Jayanth Banavar, the response "scales." This means that the almost instantaneous transmission of visual and other information (from the most sensitive individuals) to an ever-larger portion of the population follows an exponential law typical of physical phase transitions. In that sense, "collective responses" by moving populations of animals may represent another example of the dynamics which physicist Banavar and I proposed (2007) with respect to arousal systems in the brain although, of course, the two families of phenomena are related only through a type of exponential equation.

In his 2018 *Trends in Cognitive Science* review, Couzin adopts a still broader point of view by reflecting that behavior is dynamic, that it flows through time, and that group behaviors can change very rapidly. He emphasizes "criticality," a physical concept brought into neuroscience by Rockefeller University physicist Marcelo Magnasco, which describes the explosive capacity for rapid change of a system which is neither, at one extreme, sluggish or at the other extreme, chaotic and unstable. In a group of fish, birds or other animals, the capacity to turn local interactions into a group response near the group's "critical point" can produce the kind of near-synchronies of behaviors which I have been discussing.

In summary, we see how a high-level and slightly mystical concept of animal behavioral "collective cognition" can, by careful study, be explained through a combination of experimental analysis and quantitative thinking. We do not readily find this type of cognition in humans, although a social scientist might find analogies in large-scale responses of the relatively weak to a strong, dynamic leader. Such responses, now amplified by the instantaneousness of media, can sometimes lead to catastrophe.

What we do find in humans, however, is the type of social brain that evolved in fish over half a billion years ago. As I have emphasized in every chapter so far, brain mechanisms and evolution must be regarded in tandem, along with the development of babies, in order to understand the origin of human social behavior. Fish provide a stunning example of some of the earliest social behaviors in nature. In this sense, they are both distant from us (in time and habitat) but also curiously close. They allow us to contemplate certain unifying impulses across the vertebrate phylum. Though I have a lot more to discuss in coming chapters, I think I have by now made clear that sociality is embedded in nature, reinforced by eons of evolution.

There are, of course, fish (such as the lionfish) that prefer to remain alone. Among the primates, orangutans share this preference. Vertebrates such as leopards and black bears are notorious loners. Add to these, in no special order, sea turtles, moles, black rhinos, pandas, koalas, sloths, and armadillos. But these examples stand out. The social impulse, while differently expressed among different species and phyla, is basic to survival. Because most animals cannot build fortresses, they form groups. They maintain those groups through social behavior.

In the next chapter, I will discuss elemental social behaviors that characterize virtually all vertebrate species. The social behaviors that present most reliably are sex, parenting and aggression. My own lab and others have found hormonal and neuronal explanations for some of these.

References

Abril-de-Abreu, R., et al., 2015. Social eavesdropping in zebrafish: tuning of attention to social interactions. Sci. Rep. 5, 12678.

Baran, N.M., et al., 2017. Applying gene regulatory network logic to the evolution of social behavior. PNAS 114, 5886–5893.

Calderon, D.P., et al., 2016. Generalized CNS arousal: existence, mechanisms, and submission to quantitative analysis. Neurosci. Biobehav. Rev. 68, 167–176.

Cottingham, S.L., Pfaff, D.W., 1986. Interconnectedness of steroid hormonebinding neurons: existence and implications. In: Ganten, D., Pfaff, D.W. (Eds.), Current Topics of Neuroendocrinology. vol. 7, pp. 223–250.

Couzin, I.D., 2018. Synchronization: the key to effective communication in animal collectives. Trends Cogn. Sci. 22, 844–846.

Davis, R.E., et al., 1977. Autoradiographic localization of sex steroid concentrating cells in the brains of the teleost macropodus perularis paradise fish. Gen. Comp. Endocrinol. 33, 496–509.

Faustino, A.I., et al., 2017. Mechanisms of social buffering of fear in zebrafish. Sci. Rep. 7, 44329.

Fernald, R.D., 2002. Social regulation of the brain: sex, size and status. Novartis Found. Symp. 244, 169–184.

Fernald, R.D., 2012. Social control of the brain. Annu. Rev. Neurosci. 35, 133–151.

Flack, A., et al., 2018. From local collective behavior to global migratory patterns in white storks. Science 360 (6391), 911–914.

Godwin, J., Thompson, R., 2012. Nonapeptides and social behavior in fishes. Horm. Behav. 61 (3), 230–238.

Goncalves, D., et al., 2017. Neuroendocrinology of social behavior in teleost fish. In: Pfaff, D., Volkow, N. (Eds.), Hormones, Brain and Behavior, thirrd ed. vols. 2 of 5. Academic Press/Elsevier, San Diego, CA, pp. 3–19.

Goodson, J.L., 2005. The vertebrate social behavior network: evolutionary themes and variations. Horm. Behav. 48 (1), 11–22.

Hawkins, M.B., et al., 2000. Identification of third distinct estrogen receptor and reclassification of estrogen receptors in teleosts. Proc. Natl. Acad. Sci. 97, 10751–10756.

Hein, A.M., et al., 2015. The evolution of distributed sensing and collective computation in animal populations. elife 4, e10955.

Hein, A.M., et al., 2018. Conserved behavioral circuits govern high-speed decision-making in wild fish shoals. Proc. Natl. Acad. Sci. U. S. A. 115 (48), 12224–12228.

Jolles, J.W., et al., 2017. Consistent individual differences drive collective behavior and group functioning of schooling fish. Curr. Biol. 27 (18), 2862–2868.

Juntti Scott, A., et al., 2016. A neural basis for control of cichlid female reproductive behavior by prostaglandin F2α. Curr. Biol. 26, 943–949.

Kobayashi, Y., et al., 2013. Diversity and plasticity of sex determination and differentiation in fishes. Sex. Dev. 7, 115–125.

Lal, P., et al., 2018. Identification of a neuronal population in the telencephalon essential for fear conditioning in zebrafish. BMC Biol. 16, 45–55.

Liu, X., et al., 2013. Effects of TDCPP or TPP on gene transcriptions and hormones of HPG axis, and their consequences on reproduction in adult zebrafish. Aquat. Toxicol. 134, 104–111.

Liu, X., et al., 2016. Long term exposure to triphenylphosphate alters hormone balance and HPG, HPI and HPT gene expression in zebrafish. Environ. Toxicol. Chem. 35, 2288–2296.

Liu, H., et al., 2017. Sexual plasticity: a fishy tale. Mol. Reprod. Dev. 84, 171–194.

Lopes, J.S., et al., 2015. Brain transcriptomic response to social eavesdropping in zebrafish (Danio rerio). PLoS One 10 (12), e0145801. eCollection.

Luo, G.R., et al., 2008. Nr4a2 is essential for the differentiation of dopaminergic neurons during zebrafish embryogenesis. Mol. Cell. Neurosci. 39, 202–210.

Maruska, K.P., Fernald, R.D., 2018. Astatotilapia burtoni: a model system for analyzing the neurobiology of behavior. ACS Chem. Neurosci. 9 (8), 1951–1962.

Mathuru, A.S., et al., 2017. Familiarity with companions aids recovery from fear in zebrafish. bioRxiv:098509.

Nagy, M., et al., 2018. Synchronization, coordination and collective sensing during thermalling flight of freely migrating white storks. Philos. Trans. R. Soc. Lond. B. Biol. Sci. 373 (1746). pii: 20170011.

Newman, S.W., 1999. The medial extended amygdala in male reproductive behavior. A node in the mammalian social behavior network. Ann. N.Y. Acad. Sci. 877, 242–257.

O'Connell, L.A., Hofmann, H.A., 2011. The vertebrate mesolimbic reward system and social behavior network: a comparative synthesis. J. Comp. Neurol. 519 (18), 3599–3639.

O'Connell, L.A., Hofmann, H.A., 2012. Evolution of a vertebrate social decision-making network. Science 336 (6085), 1154–1157.

Oliveira, R.F., et al., 2016. Assessment of fight outcome is needed to activate socially driven transcriptional changes in the zebrafish brain. Proc. Natl. Acad. Sci. U. S. A.

Parhar, I.S., 2002. Cell migration and evolutionary significance of GnRH subtypes. Prog. Brain Res. 141, 3–17.

Parhar, I.S., et al., 1995a. Embryonic development of gonadotropinreleasing hormone neurons in the sockeye salmon. J. Comp. Neurol. 362, 256–270.

Parhar, I.S., et al., 1995b. Genes and behavior as studied through GnRH neurons: comparative and functional aspects. Cell. Mol. Neurobiol. 15, 107–115.

Parhar, I.S., et al., 1996. Gonadotropin-releasing hormone gene expression in teleosts. Mol. Brain Res. 41, 216–227.
Parhar, I.S., et al., 2016. Reproductive neuroendocrine pathways of social behavior. Front. Endocrinol. (Lausanne). 7, 28–48.
Pellegrini, E., et al., 2005. Relationships between aromatase and estrogen receptors in the brain of teleost fish. Gen. Comp. Endocrinol. 142, 60–66.
Pfaff, D.W., 2006. Braon Arousal and Information Theory. Harvard University Press, Cambridge.
Pfaff, D.W., 2017. How the Vertebrate Brain Regulates Behavior: Direct from the Lab. Harvard University Press, Cambridge, MA.
Pfaff, D.W., Banavar, J.R., 2007. A theoretical framework for CNS arousal. BioEssays 29 (8), 803–810.
Pfaff, D.W., Keiner, M., 1973. Atlas of estradiol-concentrating cells in the central nervous system of the female rat. J. Comp. Neurol. 151, 121158.
Rink, E., Wullimann, M.F., 2001. The teleostean (zebrafish) dopaminergic system ascending to the subpallium (striatum) is located in the basal diencephalon (posterior tuberculum). Brain Res. 889, 316–330.
Rosenthal, S.B., et al., 2015. Revealing the hidden networks of interaction in mobile animal groups allows prediction of complex behavioral contagion. Proc. Natl. Acad. Sci. U. S. A. 112 (15), 4690–4695.
Scales, H., 2018. Eye of the Shoal. Bloomsbury Sigma, London.
Schwanzel-Fukuda, M., Pfaff, D.W., 1989. Origin of luteinizing hormone-releasing hormone neurons. Nature. 338, 161–164.
Shubin, N., 2009. Your Inner Fish. Vintage/Random House, New York.
Strandburg-Peshkin, A., et al., 2013. Visual sensory networks and effective information transfer in animal groups. Curr. Biol. 23 (17), R709–R711.
Tabansky, I., et al., 2018. Molecular profiling of reticular gigantocellularis neurons indicates that eNOS modulates environmentally dependent levels of arousal. Proc. Nat. Acad. Sci. 115, E6900–E6909.
Teles, M.C., et al., 2015. Social interactions elicit rapid shifts in functional connectivity in the social decision-making network of zebrafish. Proc. R. Soc. Biol. Sci. 282, 1099.
Thompson, R.R., Walton, J.G., 2013. Social regulatory functions of vasotocin and isotocin in fish. In: Choleris, E., et al. (Eds.), Oxytocin, Vasopressin and Related Peptides in the Regulation of Behavior. Cambridge University Press, Cambridge, pp. 75–96.
Wircer, E., et al., 2017. Homeodomain protein Otp affects developmental neuropeptide switching in oxytocin neurons associated with a long-term effect on social behavior. Elife 6. pii: e22170.
Wullimann, M.F., Rink, E., 2002. The teleostean forebrain: a comparative and developmental view based on early proliferation, Pax6 activity and catecholaminergic organization. Brain Res. Bull. 57, 363–370.

Further reading

Fan, G., et al., 2018. Chromosome-level reference genome of the Siamese fighting fish Betta splendens, a model species for the study of aggression. Gigascience 7 (11).
Ota, Y., et al., 1999. Differences in seasonal expression of neurohypophysial hormone genes in ordinary and precocious male masu salmon. Gen. Comp. Endocrinol. 116, 40–48.
Romano, D., et al., 2017. Multiple cues produced by a robotic fish modulate aggressive behaviour in Siamese fighting fishes. Sci. Rep. 7 (1), 4667.
Yang, X., Marchetti, M.C., 2015. Hydrodynamics of turning flocks. Phys. Rev Lett. 115 (25), 258101.

CHAPTER 4

Elemental social behaviors

The most obvious social approach behaviors are those necessary for survival of the species. Sex, without which no vertebrate species can survive, is by definition social. In this chapter, I will look at sex, among other social behaviors that are so basic to a species' survival that the species would disappear were these behaviors to decline significantly.

For this chapter I am less concerned with individuals' survival, but instead concentrate on behaviors that have evolved—as species evolved—to ensure that the *species* would likely survive. These basic, elemental behaviors, are calibrated to survival; it is why they exist and why they evolved into the forms they now take. They are part of the evolutionary prong of my argument. Additionally, they depend on the brain mechanisms that constitute my argument's third prong. Finally, since parenting, like sex, is another such elemental behavior, the second prong of my argument involving babies' development also comes into play.

Besides sex and parenting, I will also discuss aggression. This is another basic social behavior even though we think of it as anti-social. It is important to note, moreover, that aggressiveness is a state of mind that can manifest in connection to sex and parenting. Animals will fight over sex partners, and parents will become aggressive if their young need food or appear to be threatened. Aggression is therefore perhaps the most complex of the three behaviors that I discuss here because the stimuli that can provoke it are so diverse.

The most primitive prosocial behaviors serve the needs of reproduction. The relevant brain mechanisms are the best worked out compared to all other behaviors, and are best analyzed in simple laboratory animals because their reproductive behaviors are straight-forward and stereotyped. Further, for the neuroscientist, regulation of their sex behaviors by steroid hormones such as estrogens and testosterone offer analytic advantages. Because the hormone receptors are ligand-activated transcription factors, we have treated behavioral questions as questions of molecular biology. We used the effects of the behaviorally effective steroid hormones on specific nerve cells in the hypothalamus to figure out how the entire behavior works.

https://doi.org/10.1016/B978-0-323-85861-8.00008-6

Nikolaas Tinbergen

In this chapter, I will analyze two types of social approaches, sexual and maternal, in terms of the framework established by Tinbergen (1963). As I note below, Tinbergen set out the requirements for stating when and how a problem in the neuroscience of behavior has been solved. For sexual behavior (Pfaff, 2017), my lab solved the problem according to his criteria some time ago. Of course, we performed our work using animals, and the behavioral patterns that we observed relate most particularly to the animals that we used. But (as I show later in this chapter) they translate to humans in broad outline, and establish that sexual behavior evolved over time to ensure that the sexes *wanted* to have sex; that males were attracted to females; and that females went to considerable lengths to attract males and signal their interest. We showed that the necessary brain mechanisms evolved, a crucial part of the story I am telling regarding the origin of social behavior.

Now that I think of it, my intertwining theory of evolution and development and mechanisms may, unconsciously, have been influenced by Tinbergen's framework, as follows.

Sex

So how is it possible to successfully analyze sex behavior regulation within the Tinbergen framework that applies to all natural social behaviors? Pfaff et al. (2019) used the Tinbergen (1963) framework, which held that a problem in the analysis of behavior is not solved until scientists understand its (i) development, (ii) mechanisms, (iii) function, and (iv) evolution. We established that sex (which everybody thinks they understand, and nobody cares to anatomize clinically) can be broken down into well-defined microbehaviors that encourage the male to mate with the female. What is amazing is how little is left to chance in what amounts to a well-coordinated sequence of action-and-reaction leading up to the actual mating. Though each action and reaction seems simple, *together* they are highly complex and reflect an evolutionary process that favored each of them in combination. Evolution had to get sex right, and had to ensure that it was adapted to survival in the individuals' environment.

Development. The simplest example of a mammalian social behavior is the mating response of female mammals: it constitutes a standing response (following rapid locomotion during courtship) coupled with vertebral dorsiflexion. Called "lordosis," it depends on circulating estrogenic hormones which must enter the brain and bind to estrogen receptor alpha in neurons in the ventromedial nucleus of the hypothalamus.

Ovaries function to secrete estrogens as early as postnatal day 4 with a significant increase starting on postnatal day 7 (Dutta et al., 2014). It is well established that Estrogen Receptor-α begins to be expressed on postnatal day15 (Gerlach et al., 1983; Friedman et al., 1983) and, indeed, lordosis behavior is shown first on postnatal day 15 (Kow et al., 2007). This neuroendocrine system develops further when estradiol, supplemented by progesterone stimulates the ovulatory luteinizing hormone (LH) surge. Thus, after puberty (days 30–36) the ovulatory surge of LH will occur late afternoon on the day of proestrus (the third of the 4-day female cycle), well-coordinated with lordosis behavior (just after dark) which permits fertilization by the male.

Mechanisms. Mechanisms for mating behavior in female laboratory mammals have been worked out (Pfaff, 2017). Females' social approach, "courtship," consists of luring males forward so that they bump into the female in the correct position for mounting, thus to stimulate lordosis. Two things to say about this simple social behavior in a laboratory animal: first, it is proof of principle that it actually is possible to explain a mammalian social behavior.

Second, even though human romantic and social behaviors certainly cannot be reduced to what laboratory animals do, an amazingly high amount of the underlying mechanisms discovered in animal brains actually do contribute to the regulation of instinctive human sexual behaviors.

Lordosis behavior is triggered by somatosensory stimuli on the flanks and rump, usually caused by the male mounting. Action potentials travel the well-studied cutaneous nerves to arrive in the dorsal horns of the spinal cord at lumbar levels. These signals ascend in spino-reticular and spino-thalamic tracts to form a spino-reticulo-spinal loop whose activity is controlled by estrogen-dependent signals from neurons expressing the Estrogen Receptor (α) in the ventromedial hypothalamic nucleus. Several genes expressed in those hypothalamic neurons have two properties: (i) their mRNA levels are increased in those neurons by estrogen treatment, and (ii) their proteins produced from those mRNAs subsequently foster lordosis behavior. The estrogen-dependent signal from the ventromedial (VM) hypothalamus activates neurons in the midbrain central gray, whose neurons in turn activate the behaviorally-relevant reticulospinal and vestibulospinal neurons. They activate the motor neurons for the axial, deep back muscles whose contraction causes lordosis.

This mating behavior, essential for fertilization and therefore for reproduction, is overdetermined in the sense that at several levels redundant mechanisms operate. For example, all the brain mechanisms are bilateral, and forward of the spinal cord the operation of the lordosis behavioral will survive unilateral damage. Further, it is not just one, but several transcriptional mechanisms which are estrogen-sensitive in the relevant VM hypothalamic neurons and involved in fostering lordosis. Several neurotransmitters impact VM hypothalamic to excite their electrical activity, again redundant. Finally, progesterone amplifies the estrogen effect.

Of course, there is a massive sex difference in the performance of this behavior. Male mice or rats—or females treated with testosterone during the perinatal period—rarely exhibit lordosis behavior.

Social approach by male laboratory animals is equally well understood. In accordance with the review by Florida State professor Elaine Hull (Hull and Rodriguez-Manzo, 2017), I can say that initial social approaches by male laboratory animals are not subtle. Males with adequate testosterone levels in the blood will almost always chase the female and sniff her. The neurotransmitter dopamine facilitates this social activity.

Laboratory hamsters initiate reproductive behaviors using a virtual multi-sensory "symphony." If and only if males have circulating testosterone, they will deposit smelly testosterone-dependent glandular marks on objects in their environment. Encountering these, females, if and only if they have high estrogen titers, deposit vaginal pheromonal marks on the ground with increasing concentrations near their burrows. The males run up that concentration gradient and emit ultrasonic calls, answered in turn by females' ultrasonic calls. This leads the male into the burrow where further sociosexual communication is conducted by touch. Thus, there are three sensory systems: olfaction (diffuse in time and space), ultrasound (specific in time and space), and then touch (local, indeed).

I note that under artificial experimental conditions, initiation of sexual behavior by male rats can be conditioned to stimuli which raise CNS arousal. Villanova professors William Crowley and Byron Ward state: "It is suggested that the momentary augmentation of nonspecific arousal can lead to the initiation of mating responses in noncopulators" (Crowley et al., 1973). Ann Clark and Edward Roy (1985), then at the University of Illinois, found evidence that males failing to initiate sexual approaches had lower sex hormone receptor expression in neurons of the preoptic area. A more recent theory (Pfaff, 2019) would assign much of the causation of initial sociosexual contact to an anterior/posterior integrated arousal system in the lower brainstem, with an emphasis on nucleus gigantocellularis neurons in the medullary reticular formation (Pfaff, 2019). Their findings are important to us because it is so obvious that humans will take measures, indeed, take risks to raise their levels of arousal and excitement in order to trigger sexual behaviors.

A paper in *Cell* from Nirao Shah's lab at Stanford (Bayless et al., 2019) shows the importance of neurons in the pathway from the ancient forebrain, including the amygdala, to the preoptic area and hypothalamus for stimulating males' approaches to females. A genetically defined subset of those neurons, cells which express enzymes which convert testosterone to estradiol, were crucial.

In summary, in Elaine Hull's view, the female laboratory animal's social participation is the more complex. If the female is not ready to ovulate, then she may fight off the male by escaping or by turning on him and boxing or biting him. If, however, she is immediately pre-ovulatory, she can tour the broad region of her burrow spreading estrogen-dependent odors, pheromones which attract the reproductively competent male. Then, running in front of the male, she can perform extremely fast starts and stops which cause the male to bump into her in the right position for mounting. In contrast, the male simply has to sense her estrogen-dependent pheromones and chase the female until he mounts.

Readiness among females is an interesting phenomenon. It has evolved so that a female, feeling that she is not ready, will signal the male and thus prevent his wasting energy in trying to mate with her. It is naïve to try to find direct analogies in human behavior… but without getting into them in any detail, I will simply point out that everyone can think of some. The point about readiness, as with all the micro-behaviors examined in the studies just cited, is that each species—be it laboratory mice or humans—has evolved behaviors that enable sex to "work" more or less efficiently. It could even be argued that humans have (through advanced medical technology) eliminated so many barriers to sexual efficiency that we are on a trend-line towards unstoppable overpopulation. In a natural setting, unmediated by technology, sexual efficiency is crucial. I have shown how a coordinated series of micro-behaviors work to ensure it insofar as possible.

Function. The adaptive function of lordosis behavior is obvious. The standing posture coupled with vertebral dorsiflexion allows the male to mount the female and deliver sperm. The redundancy of the neuronal mechanisms enables the individual to mate and produce offspring even in cases of injury or developmental problems. Of course, the function of the male's social approach is simply to allow mounting and fertilization.

Evolution. Female sexual selection is a driving force in evolution. Lordosis enables internal fertilization, making the ability to perform this behavior crucial to a female's evolutionary fitness (Futuyma, 1998; Birkhead and Pizzari, 2002). As foreseen in Pfaff et al. (2019), the discussion of sexual behaviors of female laboratory animals includes at least three lines of evolutionary thought. During evolution, lordosis behavior co-opted already-developed postural control pathways: vestibulospinal and reticulospinal (Cottingham et al., 1988). Further, the evolution of a female-typical mating behavior is neither limited to lordosis itself nor to the evolution of quadrupeds (in that some female birds have a similar standing posture and vertebral dorsiflexion). Thus, lordosis or similar behaviors likely evolved multiple times, suggesting that it conferred a fitness advantage onto the females that possessed this behavior and the males that were responsive to it. It is likely that female choice allowed for an increase in the fitness of the females and their offspring (Drickamer et al., 2000).

For humans, the simplest possible historically-stamped formulation would state that males will compete for a female's attention and that the female selects among the males. Of course, now the same formulation must be applied in much, much more complicated patterns of sexual relations.

Second, there was a clear need for internal fertilization to deal with the transition from living in water to living on land. There evolved corresponding changes in egg proteins (Litscher and Wassarman, 2007) and changes in sperm. But there are complexities; some fish use internal fertilization, suggesting that the optimal mating strategy is dependent on multiple environmental factors. This is supported by evidence that (i) internal fertilization has evolved more than once, and (ii) external fertilization has also possibly evolved from internal fertilization (Long et al., 2014). While mating strategies vary greatly among species, overall internal fertilization would be expected to increase the importance of female mate choice as a selective constraint on the reproductive fitness of males, contributing to the evolution of lordosis.

Third, the nuclear receptor Estrogen Receptor-alpha (ER-α), expressed in ventromedial hypothalamic neurons in a manner essential for female reproductive behavior, is a member of a large family of nuclear receptors. Current thinking states that the ancestral nuclear receptor had a high degree of similarity to ER-α and that the large family of steroid receptors "evolved according to a principle of minimal specificity: at each point in time, receptors evolved ligand recognition criteria that were just specific enough to parse the set of endogenous substances to which they were exposed" (Eick et al., 2012). New lines of thinking also consider relationships between the evolution of steroid hormone receptors and transposable elements (Hunter et al., 2015). For example, Testori et al. (2012) searched for transposable elements overlapping Estrogen Receptor α (ERα) binding peaks in publicly available ChIP-seq databases and found that "ERα preferentially targets a well-defined set of Transposable Elements (TEs) and that these TEs host combinations of transcriptional regulators involving several of known co-regulators of ERα." Regarding another nuclear receptor, the vitamin D receptor, PCR-based amplification of the Alu SINE from human and non-human primate genomic DNA and subsequent sequence analysis, revealed perfect structural conservation of the VDRE (Gombart et al., 2009).

Ultimately, for mammalian sexual behavior—which is the key to reproductive success—the evolutionary question boils down to the evolution of a gene regulatory network. This question has been approached recently (Baran et al., 2017), but much work along these lines remains to be done.

Libido. Given the tremendous amount of data about the reproductive behaviors of laboratory animals, what can we say about human reproductive behavior? In his 40s, the great psychoanalyst Sigmund Freud had given up his career as a neurologist and was devising psychological concepts which, by and large, are too amorphous—too unrelated to actual brain mechanisms—for neurobiologists to deal with. The one exception is the concept of *libido.* In Freud's work, this concept led a double life, straddling the border between the purely biological and the fuzzily psychological. Neurobiologists cannot yet explain all the complex psychological influences which affect a person's motivation to approach, romantically, another person. But Freud's concept of libido also had a second, physiologically deeper side. Libidinous thoughts and actions were seen by Freud to result from biological forces, i.e., physiological and chemical phenomena in the body. These we can explain.

Of course, the simplest idea would always be to assume that hormonal and neural mechanisms discovered in the brains of laboratory animals persist during evolution and operate in the human brain as well. After all, when biologically adaptive molecular and physiological mechanisms have evolved over a long period of time, there is no obvious reason to think they will easily disappear. Indeed, there is a lot of evidence to support the idea of conservation of reproductive biological phenomena from animal brain to human brain. No fewer than *22* lines of evidence (Pfaff, 1999) show that mechanisms underlying sex behaviors in laboratory animals have been conserved, i.e., retained in human brains.

First, the fragment of protein which is called Gonadotropin Releasing (GnRH). This fragment is composed of 10 amino acids and governs all things reproductive, in the brains of laboratory animals and human beings. In classical endocrine terms, GnRH synthesis by neurons just in front of the hypothalamus and its release in the pituitary gland are important because of what happens after it stimulated follicle stimulating hormone (FSH) and luteinizing hormone (LH) release from cells in the pituitary. In male animals and human males, this GnRH effect regulates spermatogenesis and testosterone synthesis. In female animals and in women, a large GnRH pulse into the pituitary causes the ovulatory surge of LH. No GnRH surge, no ovulation.

GnRH facilitates sex behavior too, in females (Pfaff, 1973) and in males (Myers and Baum, 1980). Most unusual, as noted above, GnRH synthesizing neurons are not born directly in the brain like other neurons (Fig. 1 in Chapter 3). Instead, they are born in the nose, in a cell layer called the olfactory placode (Schwanzel-Fukuda and Pfaff, 1989). From there they migrate early in life into the brain. Interruptions in the GnRH neuronal migration from the nose into the brain are associated with a loss of libido in human patients (Schwanzel-Fukuda et al., 1989). This migration turns out to be a broad biological phenomenon—it happens in fish (Parhar et al., 1995, 1996) and in humans, thus from "the fish to the philosopher."

My main point is that the most fundamental aspects of GnRH chemistry, cell biology, and functions in the body have remained the same among mammals including humans. Thus, evolutionary conservation of one set of mechanisms is related to the primitive, physiological components of libido. Accordingly, we can see how GnRH chemistry and related functions, intrinsic to human libido, fits into the evolutionary and brain mechanism prongs of my basic argument. Human sociality, manifested in the fundamental chemistry/behaviors associated with sex, is in this respect the product of the evolution of brain mechanisms that ensure (insofar as possible) the species' success. You cannot think of sex and the sex-drive *except* in evolutionary terms since humans would not now exist but for the evolution of successful sexual behaviors as these are regulated by the relevant brain mechanisms. It is ironic that while we tend to think of sexual attraction as immediate ("I knew in the first ten seconds"), it is also an evolutionary development that embodies eons of adaptation. It is safe to say that the one ("I knew") was ensured by the other.

In a larger sense, we can see the importance to human sociality of evolutionary conservation. It is not just that our brain mechanisms for sex evolved to ensure our survival, but that once evolution hit on a successful modus operandi it found its way into many subsequent animal species.

I pointed out similar developmental patterns in the chapter on fish. These trans-species similarities, in various contexts, simply emphasize the impact of evolution on brain mechanisms. Human sociality is part of a much larger evolutionary pattern that favors—that facilitates—sociality. Thus, at the risk of sounding like a broken record, I want to emphasize, yet again, that we cannot understand the origin of human sociality without ranging back in time and across species. When I get to another subject of this chapter, parenting, it will become clear that in order to understand our sociality we must also range from the tiniest molecules that constitute our brain chemistry to parental behavior and babies' developmental response.

Genes and human sex behavior. The GnRH story shows how gene expression in the brain can be related to human social behavior but, at the same time, illustrates the complexity and multiplicity of the causal route. In men, Kallmann Syndrome patients' loss of libido can be laid to one or more genetic mutations (Taroc et al., 2019; Crowley Jr., 2017). These patients have low libido *because* the affected individuals have very low testosterone levels, *because* not enough luteinizing hormone came from the pituitary gland, *because* not enough GnRH reached the pituitary coming from the brain *because* during development GnRH neurons did not reach the hypothalamus, *because* of a failure to migrate from the nose into the brain, *because* of mutations in the Kallmann's gene. Thus, while the gene/social behavior linkage in humans is established, it works through six causal/sequential steps.

Further, we should consider the relationship of hormones to sex. The sex steroid hormones are chemically identical in laboratory animals and humans. Consider, for example, estrogenic hormones like estradiol and estrone. They are essential for normal reproductive behaviors in female laboratory rats and mice, and also support reproductive performance in women. Likewise, testosterone, which fosters sex behaviors in male animals and in men. The tricky part is that an enzyme called aromatase chemically transforms testosterone into estradiol and that, in fact, this chemical conversion is necessary for testosterone to be effective in the cases of some behaviors. The enzyme aromatase is important in the human brain as well.

The genes that code for the proteins considered to be "hormone receptors" are also largely conserved from mouse and rat brains to human brains. Hormone receptor chemistry remains the same. Additionally, the neuroanatomical distribution and connections of cells that have these sex hormone receptors—a system of cells in the ancient forebrain and hypothalamus—remains the same in the human brain. In fact, hypothalamic neuroanatomy in general is conserved.

Consider the large family of major molecular and biological phenomena that are also conserved. The fundamental structure of DNA, the organization of gene enhancers, gene promoters and coding regions of genes important for social behaviors that we study in animal brains all remain in the human brain. The molecular biology of neurons, as well as the chemistry of neurotransmitters and neuropeptides ("neuromodulators") in the human brain were not just invented suddenly for the support of human libido. By and large, the neuroanatomy and molecular regulation of these substances are no different between lab animals and humans.

Likewise, the mechanisms regulating neural signaling by electrical action potentials and the resultant deposition of neurotransmitters at synapses are fundamentally similar among mammalian species. Synaptic mechanisms appear identical between lab animals and humans. The pharmacology of brain mechanisms regulating sociosexual behaviors has never been shown to be different in humans.

Obviously, the basic requirements for fertilization remain in place as we move from lab animals to human sexuality and libido. In addition, the biology of sex differences (of importance only for the nitty gritty of reproduction) remain in place.

Look back in this chapter and count them. Across more than 20 axes of sociosexual mechanisms, laboratory animal chemistries and physiology are conserved into the human brain. Thus, a tremendous amount of evidence tells us that certain defined steps that cause male and female animals to approach each other also occur in the human brain. It is as if evolution, having established mechanisms that succeeded over time, simply kept them in order to ensure that a diversity of species could avail themselves of what worked. It's like the old joke that says "if it ain't broke, don't fix it." In evolution, however, this is no joke.

Cultural contributions to variations in human sexuality. While the best studied mechanisms of laboratory animal reproductive and parental behaviors have yielded solutions which are scientifically satisfying, there is a real question about how far they go in explaining some aspects of human instinctual behaviors.

As soon as we move from lower animal sexuality to human sexual attraction which leads to social approach, certainty disappears. It comes as no surprise that the variety, subtlety, and cognitive content of human reproductive instincts go immeasurably further than what we see in the lab animals whose brain mechanisms we have worked out. Consider that Facebook has no less than *50* designations one can use to describe one's sexual identity.

In the United Kingdom, that number has been *71*. Accordingly, this book does not even try to take on the variety and depth of the human sexual experience. In fact, any such enquiry would be extraneous to my argument, which seeks to explain only the deepest scientific basis for why we initiate social contact.

Psychiatrists and psychologists emphasize tremendous individual variability in the relationship between the motivational sexual force and subsequent romantic behaviors. And social rules change rapidly. University of Washington's David Barash's (2018, p. 182) way of describing this situation refers to "the vast gulf that has opened up between our biological selves -- changing very slowly over vast expanses of time -- and our cultural selves..." for whom "the rate of changes has been changing, accelerating." Under these circumstances, any scientific explanation of the first spark of romantic interest may meet push-back from cultural critics. Yet without seeking to draw some fictitious line between our biological and cultural "selves," I think there are still certain very basic determinants, introduced above, of human social approach that are not changing and cannot be willed away.

Infidelity is a problem psychologists have faced. That is, once we commit to an attachment and get married, either party can become interested in someone else. When asked whether we are disinclined towards monogamy, psychologist and evolutionary biologist David Barash, at the University of Washington, answered with a definite "yes" (Barash, 2018). His assessment of the data indicates that we are not monogamous. He observes that even "a quick review of the evidence [shows] that polygamy is the default human system" (p. 135). Men engage in polygyny, women in polyandry. But this morals-defying behavior just underlines my point: *sex persists, no matter what.* It is such a basic impulse that we do not need any cultural approval to throw it into stark relief, i.e., to demonstrate how baked into human conduct it is. That is, the prejudice against infidelity is moral—a social construction—and has nothing to do with the continuing attraction of one sex to the other (or, as I suggest below, the attraction of one differently oriented person to another, compatible person). Even when sex is available on the other side of the bed, some go looking for more. Interest in sex is so great that when sex is not available, people masturbate, consume porn, have fantasies. Though sex may seem to be everywhere—it sells cars, shaving cream, underwear—where it is not, we contrive it as best we can. So, my point is that no matter how we slice up the social construction of sex and gender, some form of initial sexual attraction is a natural form of the fundamental human social impulse. And this book is devoted to explaining this fundamental human social impulse.

Clearly while "attraction" usually was formulated in terms of heterosexuality, it applies (as suggested above) to all shades and variations of sexuality as they are now openly avowed and practiced. Gay men and lesbians, for example, feel as strongly attracted as do heterosexuals. The same can be said for bisexuals and members of other sexual orientations. Whether or not you believe that these groups are mere social constructions, the fact is that their members experience sexual attraction as intensely as anyone else.

A deep neurobiological issue related to the phrase "sexual attraction" is how we are able to recognize faces (see Chapter 5). If you know nothing else about the science of facial recognition, you know from online dating that a face can be instantly attractive or an instant turn-off. If you don't like someone's face, you won't read the profile that they posted. So how *do* we recognize faces?

As I will show in Chapter 5, neuroscientists have for a long time wondered how we would build our understanding of the visual recognition of all sorts of objects. The pioneer neurophysiologist, Steven Kuffler, had provided one approach during his studies of the retina. Nobel prize winners David Hubel and Torsten Wiesel, at Harvard Medical School, had shown how cells in the primary visual cortex code their electrical activity according to lines, angles, corners, and other dimensional indicia. But what about the recognition of objects that are important to us, e.g., because they relate to people whom we care about? Neurophysiologists joked about finding a neuron tuned specifically to fire electrical discharges only when we looked at our grandmother—a "grandmother cell".

Then, Princeton neurophysiologist Charles Gross, recording from neurons in the inferotemporal lobe of a monkey, discovered neurons which would fire action potentials only when the monkey was looking at a face. Following that up with humans, MIT neuroscientist Nancy Kanwisher found spots in the human cerebral cortex which responded to faces. Going even further, Winrich Freiwald at Rockefeller discovered nerve cell groups in the frontal cortex whose activation is specifically tied to the sight of social interactions (Sliwa and Freiwald, 2017) (Fig. 1). I will discuss all of this in the next chapter. I note here, however, that in a recent review by Ralph Adolphs' lab (2017), he asks "how looking at people's faces can influence prosocial behaviors towards them." Of course, physically attractive people are by definition people whom we like and want to approach. So, to the extent that looking at faces raises our sexual interest, it can influence prosocial

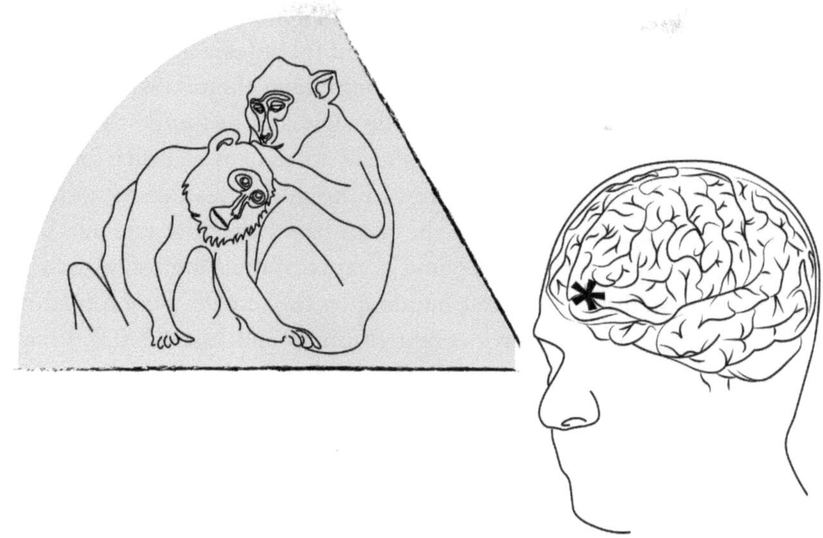

Fig. 1 Certain neurons in the monkey forebrain respond specifically to the sight of social encounters.

behavior (which we kind of knew anyway if we have ever looked for a date online). With regard to facial recognition, Adolphs and his colleagues link activity in several brain areas such as the amygdala, the insula and the prefrontal cortex.

So, modern neurobiology speaks powerfully at least to how that initial spark of attraction is lit up.

Parental behaviors

In terms of Tinbergen's four criteria for establishing that we understand a behavior, what about parental approaches to their young? We know less about parental compared to sexual behaviors because parental behaviors have more complicated hormonal concomitants and sensory-motor topographies than female reproductive behavior. However, we have the largest amount of data for maternal behaviors of laboratory animals such as rats and mice.

Development. Among mammals, maternal behaviors can begin as soon as the female can support a pregnancy and normal delivery, but "primiparous" mothers (females giving birth for the first time) are more susceptible to interruption of their parental behaviors than females with previous experience. Nursing is the most universally expressed maternal behavior among mammalian females (González-Mariscal et al., 2017).

The first thing pregnant female laboratory animals do to prepare for contact with pups is to build nests, important for at least three reasons. First, when the pups are born, the nest will help keep them warm. Second, it helps to hide newborn pups from outside view and to muffle their sounds. Third, it provides the place for the female to remain stationary and safe, and thus able to nurse the pups—stress to the mother can disrupt maternal behavior. Neuroscientists measure the nests based on their size and ability to conceal the pups from outside view because of their high sides. Sex hormones are required for good nest building by the female: a combination of estrogens and progesterone works best (Numan and Insel, 2003). In addition to steroid hormones, oxytocin clearly plays an important role, since an oxytocin receptor antagonist or an antibody raised against oxytocin interferes with the ability of estrogen treatment to support nest construction (Fahrbach et al., 1984, 1985).

Another early sign of social attention to the pups is that when scattered they should be retrieved to the nest. Estrogen treatment triples the frequency of retrieving. Treating the estrogen-injected female with an oxytocin receptor antagonist or an antibody raised against oxytocin reduced retrieving to levels at or below those achieved without any estrogen treatment at all (Fahrbach et al., 1984, 1985).

Mechanisms. Classical work by Numan and others (Numan and Insel, 2003) established preoptic area neurons as required for maternal behavior. In a step towards unraveling the transcriptional steps required, it was found that to support normal maternal behavior, the ligand-activated transcription factor estrogen receptor alpha (ER-α) needed to be expressed in these preoptic neurons (Ribeiro et al., 2012; Ogawa et al., 1998). The full scope of maternal behavior—building the nest, retrieving pups to the nest, licking and warming the pups and nursing are facilitated by the neuropeptide oxytocin and transcription of the Estrogen-dependent oxytocin receptor. Some recent molecular evidence has implicated MPOA neurons which express galanin (Wu et al., 2014). Most interesting are the data showing not only that preoptic neurons expressing ER-α are necessary and sufficient for elements of maternal behavior, but also that they are electrically activated during pup retrieval and drive the behavior in part by connections to a part of the midbrain important for behavioral reward (Fang et al., 2018). That is, Yi-Ya Fang, working in the lab of Dayu Lin at New York University (Fang et al., 2018) used a large battery of current genetic and molecular approaches to build on classical understanding of exactly how the social aspects of maternal behaviors work. Social approaches to the pups by the female depend absolutely on nerve cells in the medial preoptic area which express the primary estrogen receptor. This finding used new methodologies to replicate the antisense DNA finding from my lab: reducing estrogen receptor expression in medial preoptic neurons essentially abolished maternal behaviors (Ribeiro et al., 2012). "Reversible inactivation of $MPOA^{Esr1+}$ cells impares those behaviors, whereas optogenetic activation induces immediate pup retrieval." Indeed, Fang and Lin showed that such preoptic area neurons become active during in vivo recordings to demonstrate "preferential activation of $MPOA^{Esr1+}$ cells during maternal behavior."

How does this social command get out of the medial preoptic area to activate appropriate motoric responses to the pups? The most understandable answer to this question involved movement- and reward-associated ventral tegmental area (VTA) non-dopaminergic cells in the ventral midbrain. These dopaminergic cells were reliably activated during social approach to the pups. Fang's and Lin's conclusions were in close accord with the findings of Numan and Stolzenberg (2009), who showed that dopaminergic actions on preoptic neurons then stimulate the mesolimbic dopamine system to facilitate social approaches to the pups. Putting this all together, we can say that preoptic neurons have the molecular properties to add to the estrogen dependence of maternal behaviors, but that the full transcriptional consequences of estrogen receptor occupancy have not yet explored. Outputs of preoptic neurons to the ventral tegmental area could explain initiation of movements towards pup ultrasounds and perhaps even add reward value to pup retrieval.

Thus, in terms of the instinct towards social approach, Yi-Ya Fang's and others' work demonstrates the essential role of estrogen receptor-expressing preoptic area neurons in governing the female's approach to the young. But in addition, recent data show the importance of inputs from the ancient forebrain to the preoptic area and hypothalamus. A recent paper in *Cell* (Chen et al., 2019) showed that a genetically defined subset of neurons in the medial amygdala "promotes parental behavior in females."

The nine amino acid neuropeptide oxytocin is central to mechanisms underlying maternal behavior. Having demonstrated the importance of oxytocin (Fahrbach et al., 1984, 1985) and having studied oxytocin gene expression using the histochemical method called in situ hybridization, members of my lab were keen to study gene expression for its receptor in the brain. Estrogen administration turns on expression of the gene for the oxytocin receptor (Chung et al., 1991; Quiñones-Jenab et al., 1997) through molecular mechanisms which we analyzed (Vasudevan et al., 2001;

Gagnidze et al., 2013; Faustino et al., 2015) with import for social behaviors. For example, oxytocin gene knock-out mice were aggressive when studied in a semi-natural environment (Ragnauth et al., 2005) and the use of antisense DNA directed against oxytocin receptor messenger RNA (McCarthy et al., 1994) knocked down sexual behavior. Elena Choleris, a leading postdoctoral researcher in my lab, now professor at Guelph, in Canada, conceived of a "four gene micronet" essential for social recognition (Choleris et al., 2003, 2006, 2007). Those four genes express estrogen receptors ER-α and ER-β, oxytocin and the oxytocin receptor; and their activities show the importance of coordinated functions of the hypothalamus and amygdala.

Primates. Hormonal support for the initial responses of non-human primates towards babies is surprisingly similar to earlier findings in traditional laboratory animals such as hamsters, rats, and mice. The pregnancy hormones of squirrel monkeys were associated with 80% carrying of an infant (experimentally introduced) compared with 20% non-pregnant females (reviewed, Numan and Insel, 2003). Similar results obtain if scientists record the approach to and handling of infants by pigtail macaque monkeys. In fact, common marmoset monkeys, supported by high estrogen levels, would even learn an arbitrary behavioral response, a bar press, to gain access to marmoset infant stimuli, indicating that social approach responses and such stimuli were rewarding (see Numan and Insel, Fig. 2.5). Since, during late pregnancy, progesterone levels may decline while estrogens are still rising, the estrogen-to-progesterone ratio is often cited as the key hormonal parameter.

Thus, all of these results in common laboratory animals and in primates tell us that steroid sex hormones and the neuropeptide oxytocin support the initial social approaches to infants, whether those involve nest building, retrieving, carrying, or even performing arbitrarily chosen behavioral responses.

Function. Clearly, the function of mammalian parental behavior is to enable the young to survive until the age of reproductive competence, thus prolonging the species' survival. In the last analysis, of course, all parental behaviors have an immediate function—e.g., making sure that the young are warm and fed—and an ultimate function of ensuring the species' survival. When we consider parenting as a social impulse, these two functional levels cannot be separated. That is, the social impulse that parents feel towards their young (and vice versa) evolved to ensure the survival of the parents' babies and—through those babies—that of the species.

Evolution. The evolution of maternal behavior more or less defines the evolution of mammals, and therefore is no less complex than the evolution of mammals. Recently, it has been considered in the light of oxytocin's powerful influences (Carter, 2018), the onset of maternal behavior after adolescence (Sisk, 2016), and human attachment (Feldman, 2017).

No author has better captured the complexities of human and nonhuman primate maternal behaviors than the renowned anthropologist Sarah Blaffer Hrdy (1999) at the University of California, Davis. Far from the old stereotype of the "selfless" woman, obeying "maternal instincts," Hrdy's picture leads to an understanding of ambitious, complex women, "a multifaceted view of mothers, featuring flexible actors whose responses are contingent on circumstances" (p. 535).

Further, Hrdy (2009) developed the subject of so-called "alloparental behavior," in which females who were not the mothers of a particular infant would share in the care of that infant. The female non-human primate, not the mother, would approach that infant, touch it, carry it, protect it. The biologically adaptive and hence evolutionary value of alloparental behavior has been vigorously debated, with the obvious expectation that under the environmental conditions where it appears for a given species, alloparental behavior enhances the survival of that species. (reviewed, Numan and Insel, 2003). The famous claim that "it takes a village" is consistent with Hrdy's argument.

Alloparenting occurs among human mothers as well. In certain tribes it is shown to depend on kinship and reciprocity—women are more likely to care for children of other women who themselves served in this way. In complex modern human societies, alloparenting can take multiple and subtle forms, mainly based on sharing information, e.g., about breastfeeding, formulas, age-appropriate toys, and other concerns (including, lately, dealing with coronavirus). Needless to say, there is a vast array of websites and mother-to-mother blogs devoted to information-sharing. Sometimes, the website is the alloparent. Such sites have names like Moms.com, AlphaMom, Cafemom, and ForEveryMom. The blogs have names that invite responsiveness: Mom Blog Society, Cool Mom Picks, and A Mom's Take. They are invariably run by women with children, offering a forum to exchange maternal tips (or just a safe space to vent).

As I have said so often in this book, these sites represent the technological extension of a traditional practice. They offer alloparenting by other means but, in addition, they constitute a further evolution of the practice that expands the practice beyond the purely human. Once a Mom sets up a blog, she is not just another mother offering help, and the mother who signs in is not just the recipient of that help. Both are tied together (even as the relationship is also mediated) by technology. This has changed the practice of alloparenting (or bent its arc), likely forever. Indeed, it has changed parenting. Mothers have always exchanged advice, but this is a virtual exchange in a community that is potentially infinite. Mothers anywhere can be part of the same community, and can access cultural practices they never even imagined existed. This is evolution on an entirely new scale—the speed of its development is awesome, and its reach is incalculable. Maternal instincts, emanating from brain mechanisms, are evolving to be continuous with techno-moms in cyberspace.

Will individual mothers' instincts somehow be repressed or warped given the constant access to remote alloparents? When there is no need to rely on one's own instincts because there is always a massive community to provide advice and guidance, does that stifle maternal instincts, or throw them into doubt? Does it mean that we mothers may be more likely to trust a remote alloparent than, say, someone in the community? Only the continued evolution of motherhood (as it further integrates with technology) will provide some answers.

Speaking from a more traditional evolutionary perspective, Hrdy generalizes from her studies of parental and alloparental behavior to envision "the evolutionary origins of mutual understanding." She notes (p. 42) that "highly intuitive moms not only perceive what irks their babies, a skill that enhances their ability to care for them, but also are better equipped to guide immatures as they acquire survival-enhancing skills. Mind reading is advantageous to mothers." Equipped with "empathic awareness" they are better able to understand others' intentions, an ability now enshrined in psychologists' "theory of mind" (see Chapter 9). From an evolutionary point of view, Hrdy thinks (p. 66) that such "emotionally modern humans date back to bipedal apes born with giving impulses and empathic intersubjective aptitudes"—people preadapted to get along with one another even under suboptimal and trying circumstances. In effect, Hrdy is making the claim that motherhood prepares females for acting empathetically in the world at large. If this is in fact the case, then motherhood is a vector for human sociality shared by a vast population of women. While there is no way to validate this claim empirically, it is still consistent with the fact that the social impulse is deeply embedded in human evolution.

Maestripieri and his colleagues (Saltzman and Maestripieri, 2011) confirmed the supportive roles of hormones in facilitating primate maternal behaviors. Different from simple sexual approach behaviors, steroid hormones are not absolutely necessary for maternal behaviors, but influence the level of motivation for the social approaches involved in caring for infants. Obviously, estrogens and progestins are involved, but also the protein hormone helps, as does the steroid hormone cortisol and the neuropeptide oxytocin. This makes a complex of five hormones with different time envelopes of secretion and behavioral effects. Maestripieri (2009) extended his work to free ranging macaque monkeys on Cayo Santiago, Puerto Rico, and his main results emphasized roles for the steroid hormone cortisol, as noted above.

Master scientist Alison Fleming and her team at the University of Toronto (Lonstein et al., 2015) not only reviewed studies of human maternal behavior but also explained some of the difficulties of extrapolating findings from experimental animals to humans. How important are social behaviors by the mothers towards the pups? According to Fleming, "Maternal interactions with young occupy most of the reproductive period for female mammals and are absolutely essential for offspring survival and development." Of course, mothers must attend to the young in order to allow their babies to eat, sleep and be safe.

Commonalities across cultures, for example in approaching infants in order to nurse, encouraged Fleming to think about general rules of social approach with regard to maternal behavior. As Lonstein et al. suggest, "all mothers must develop attraction to the young." Cues from the infant are important for attracting maternal approach. That is, "despite major species differences in the formation of bonds with infants, with humans forming very strong but not necessarily exclusive bonds with their infants, *there are still considerable universals between non-human and human animals in the importance of infant cues in regulating maternal behavior.* Human mothers receive multisensory stimulation right after delivery while they hold, look at and talk to their infants." Vision, odors and touch all play their roles. Auditory cues are crucial: "The most common response to a crying infant is to pick it up and hold it, which is as effective as feeding to terminate crying."

What I want to emphasize is that the cross-cultural and cross-species commonalities, often based in sense perception, demonstrate the universality of the social impulse embodied in motherhood. Such universality reflects on how evolution operates through the conservation of mechanisms that successfully support the survival of species. In this light, parenting uniquely combines the three prongs of my argument—evolution, brain mechanisms, and babies' development all at work in the origin of human sociality.

Aggression

Here I focus on male-typical aggression, both for the depth of its scientific analyses so far, and because of its importance as a problem in human society. For this account I rely, in part, on the scholarship of Brian Trainor (U Cal., Davis), Cheryl Sisk (Michigan State) and a long-standing thought leader in aggression research Randy Nelson (Ohio State and West Virginian Medical School) (Trainor et al., 2017). Following, now, the Tinbergen framework (see above), I start with the development of aggression during youth and adolescence.

Development. Of course, the big story is the rise of testosterone concentrations in the blood during adolescence. Equally interesting in the relationship between pre-adolescent play and later, more serious aggression. About the latter question, some scholars think that pre-puberal play fighting by young animals or little boys is a true precursor of adult aggression (Trainor et al., 2017), others not. In any case, according to Randy Nelson's overall summaries, "the sexual differentiation of play fighting and adult aggression is a two stage process involving steroid sex hormone actions in perinatal (or in humans, prenatal) periods and during puberty.

Mechanisms. It will come as no surprise to you that nerve cells essential for the production (and inhibition) of aggression can be found in the primitive vertebrate social brain, as introduced in Chapter 3. For example, two types of mechanisms for aggression are launched from the hypothalamus. A former priest, John Flynn became a professor at Yale Medical School and studied the effects of hypothalamic electrical stimulation in cats. If he stimulated in the ventromedial nucleus in the hypothalamus, the stimulation immediately elicited an enraged, vicious attack. If, instead, he stimulated in the lateral hypothalamus, the cat engaged in a cool, calculated prowling type of aggression as though the cat was stalking prey. So here are two sets of mechanisms which *produce* aggression.

What about inhibiting aggression? Such inhibitory mechanism must include neurons in a different part of the vertebrate social brain, the lateral septum. Damaging those neurons causes a phenomenon called "Septal Rage", in which the experimental animal will attack anything within reach. Put a pencil tip just a bit through the septal-damaged rat's cage and the rat will bite the pencil viciously. In one experiment with 12 septally damaged male rats I handled the first 11 very carefully. Then, the day before a lecture trip to Budapest, I got overconfident and the twelfth rat bit my hand terribly, and I went to Budapest with a serious bandage.

Neurochemical foundations of aggression regulation were tabulated by Trainor et al. (2017). Especially impressive for the production of aggression were those biochemicals whose presence promoted aggression and whose absence reduced impression. These included, for example, were the neurotransmitters histamine and dopamine and the neuropeptides vasopressin. Equally impressive were the biochemicals whose effect was exactly the opposite: the neurotransmitters GABA and adenosine. Hormones are in this game as well. Testosterone working through the androgen receptor and estrogens working through the estrogen receptor-alpha (Ogawa et al., 1997) are important. Interestingly, the enzyme called "aromatase" for converting androgenic hormones into estrogens is important for aggression, since when androgens get into the brain some of their behavioral effects depend on conversion into estrogens.

Genetics. Years ago my lab showed that we could make causal connections between the expression of specific genes and aggressive behaviors. Having discovered hormone receptors in the brain, I was excited to chart the behavioral consequences of knocking out genes coding for these receptors. The connections between these genes and aggressive behaviors are different between males and females. For example, Sonoko Ogawa in my lab (now a professor at "science city," Tsukuba University, Japan) discovered that knocking out the gene for an estrogen receptor *raised* aggressive behaviors by females, but *lowered* aggressive behaviors by males. Next, Masayoshi Nomura in my lab studied a different nuclear hormone receptor and discovered that knocking out this gene in males caused especially high aggression, but only just after puberty. In a provocative way our finding maps onto tendencies of human males, for whom rates of murders go up between the ages of 13 and 30.

It goes without saying that anabolic steroids, used by some (rule-breaking) athletes to stimulate the development and growth of muscle tissue, and which also have some legitimate medical uses, stimulate aggression in some instances. Not all anabolic steroids and not in all experimental conditions do this, but anabolic steroids like methyl-testosterone heighten aggression in various forms, from violence against women to verbal aggression.

An obvious genetic effect: testosterone elevates expression of the vasopressin gene, presumably one of the causal routes by which testosterone elevates aggression.

As summarized in Pfaff (2016), our work with gene knockouts led to David Anderson's work at the California Institute of Technology. His team used viral vectors and carefully directed electrical stimulation in the ventromedial nucleus of the hypothalamus to elicit attacks upon an intruder. His studies revealed "overlapping but distinct neuronal subpopulations involved in fighting and mating." This is important because of the need for males of some species to defend territory in order to attract females for mating.

Further, in addition to his intellectual leadership in this field, Randy Nelson's lab came up with quite an unusual finding: Instead of studying genetic controls of behavior by regular, conventional neurotransmitters or the kinds of hormonal studies mentioned above, he experimented with a transmitter which is actually a gas: nitric oxide. When he knocked out the gene that produces nitric oxide, he produced animals which attack very fast and very often—three to four times more aggressive than wildtype genetic controls. Likewise, when he treated animals with a drug which inhibits the nitric oxide-producing enzyme, those mice were extremely aggressive.

In sum, genetic, endocrine and neurobiological techniques have been combined to reveal, to date, some of the mechanisms underlying male-typical aggression.

Function. The usual summary would say that males aggress (a) to control resources such as territory and/or food and/or water; and (b) to preserve access to mates. And that (a) is in the service of (b). But Trainor et al. add antipredator and defensive aggression and dominance aggression.

Evolution. As you can imagine, the evolution of male-typical aggression is not recent and not even limited to vertebrates. Every species in which territory is involved in mating will likely display territory-related rituals, which often serve to prevent males from killing each other.

Perspective

The idea of including such elemental social behaviors as sex, parenting and aggression in a book about social approach came from the fact that these behaviors are *necessarily* social and therefore must depend on motivation for social approach. I had a running start because my lab had solved the mechanisms for a simple social behavior (Pfaff, 2017), and because knowledge of maternal behaviors is much advanced, evident in the work of Michael Numan and Alison Fleming. Moreover, sex hormone actions on brain and behavior have proven to be the most fertile ground for linking behavioral biology to molecular biology. But at every turn in writing this book, I realized—as I did here—that one cannot write about even the most elemental social behaviors without addressing them in context with the latest technology.

Social media and social behavior are linked forever and are evolving together. There is no escaping that. While I have not formally included technology as a fourth prong of my theory, I see it as a potential component of the evolutionary prong. My discussion of mothering and alloparenting makes this type of evolution abundantly clear. Evolution is not in the "past," but is continuing by other means and in new, uncharted directions. It is affecting the other prongs of my theory in novel ways. Science is catching up with science fiction, and we have to be clear about that. What happens in the lab under controlled conditions is still hugely important, but it is not the whole story.

In this chapter, I sought to establish how sex and parenting and aggression, when viewed together, offer a small but complete picture of how the three prongs of my argument work together in explaining how fundamental social impulses apply to natural, instinctive behaviors. However, these primitive behaviors offer *only* a small picture, a tiny percentage of the social behaviors shown by primates, including humans. The next two chapters will address complex sensory neurophysiology which allows our regulation of human social behavior to operate at more sophisticated levels of perception. Chapter 5 offers examples from the visual system. Chapter 6, from audition. And for us, of course, vision and audition usually work together.

References

Baran, N.M., McGrath, P.T., Streelman, J.T., 2017. Applying gene regulatory network logic to the evolution of social behavior. Proc. Natl. Acad. Sci. 114 (23), 5886–5893.

Barash, D.P., 2018. Through a Glass Brightly. Oxford University Press, Oxford.

Bayless, D.W., et al., 2019. Limbic neurons shape sex recognition and social behavior byin sexually naïve males. Cell 176, 1190–1205.

Carter, S.C., 2018. Oxytocin and human evolution BT. In: Hurlemann, R., Grinevich, V. (Eds.), Behavioral Pharmacology of Neuropeptides: Oxytocin. Springer International Publishing, Cham, pp. 291–319.

Chen, P.B., et al., 2019. Sexually dimorphic control of parenting behavior by the medial amygdala. Cell 176, 1206–1213.

Choleris, E., Gustafsson, J.A., et al., 2003. An estrogen-dependent four-gene micronet regulating social recognition: a study with oxytocin and estrogen receptor-alpha and -beta knockout mice. Proc. Natl. Acad. Sci. 100 (10), 6192–6197.

Choleris, E., et al., 2006. Involvement of estrogen receptor alpha, beta and oxytocin in social discrimination: a detailed behavioral analysis with knockout female mice. Genes Brain Behav. 5 (7), 528–539.

Choleris, E., Little, S.R., Mong, J.A., Puram, S.V., Langer, R., Pfaff, D.W., 2007. Microparticle-based delivery of oxytocin receptor antisense DNA in the medial amygdala blocks social recognition in female mice. Proc. Natl. Acad. Sci. 104 (11), 4670–4675.

Chung, S.R., McCabe, J.T., Pfaff, D.W., 1991. Estrogen influences on oxytocin mRNA expression in preoptic and anterior hypothalamic regions studied by in situ hybridization. J. Comp. Neurol. 307, 281–295.

Fahrbach, S., Morrell, J.I., Pfaff, D.W., 1984. Oxytocin induction of short-latency maternal behavior in nulliparous, estrogen-primed female rats. Horm. Behav. 18, 267–286.
Fahrbach, S.E., Morrell, J.I., Pfaff, D.W., 1985. Possible role for endogenous oxytocin in estrogen-facilitated maternal behavior in rats. Neuroendocrinology 40, 526–532.
Fang, Y.-Y., et al., 2018. A hypothalamic midbrain pathway essential for driving maternal behaviors. Neuron 98 (1), 192–207.
Faustino, L.C., et al., 2015. Impact of thyroid hormones on estrogen receptor-alpha-dependent transcriptional mechanisms in ventromedial hypothalamus and preoptic area. Neuroendocrinology 101, 331–346.
Feldman, R., 2017. The neurobiology of human attachments. Trends Cogn. Sci. 21 (2), 80–99.
Gagnidze, K., et al., 2013. Early histone modifications in the ventromedial hypothalamus and preoptic area following oestradiol administration. J. Neuroendocrinol. 25 (10), 939–955.
González-Mariscal, G., Caba, M., Hoffman, K., Melo, A., 2017. Parental behavior. In: Hormones, Brain and Behavior, third ed, pp. 83–116.
Hrdy, S.H., 1999. Mother Nature. Ballantine, New York.
Hrdy, S.H., 2009. Mothers and Others. Harvard University Press, Cambridge, MA.
Hull, E., Rodriguez-Manzo, G., 2017. Male sexual behavior. In: Pfaff, D., Joels, M. (Eds.), Hormones, Brain and Behavior. Academic Press (Elsevier), San Diego, CA, pp. 3–59.
Kow, L.-M., Bogun, M., Zhang, Q., Pfaff, D.W., 2007. Hormonal induction of lordosis and ear wiggling in rat pups: gender and age differences. Endocrine 32, 287–296.
Lonstein, J.S., Lévy, F., Fleming, A.S., 2015. Common and divergent psychobiological mechanisms underlying maternal behaviors in non-human and human mammals. Horm. Behav. 73, 156–185.
Maestripieri, D., et al., 2009. Mother-infant interactions in free-ranging rhesus macaques: relationships between physiological and behavioral variables. Physiol. Behav. 96 (4–5), 613–619.
McCarthy, M.M., Kleopoulos, S.P., Mobbs, C.V., Pfaff, D.W., 1994. Infusion of antisense oligodeoxynucleotides to the oxytocin receptor in the ventromedial hypothalamus reduces estrogen-induced sexual receptivity and oxytocin receptor binding in the female rat. Neuroendocrinology 59, 432–440.
Myers, B.M., Baum, M.J., 1980. Facilitation of copulatory performance in male rats by naloxone: effects of hypophysectomy, 17 alpha-estradiol, and luteinizing hormone releasing hormone. Pharmacol. Biochem. Behav. 12 (3), 365–370.
Numan, M., Insel, T.R., 2003. The neurobiology of parental behavior. In: Hormones, Brain, and Behavior. vol. 1. Springer, Heidelberg.
Numan, M., Stolzenberg, A., 2009. Medial preoptic interactions with dopamine neural systems in the control of maternal behaviors. Front. Neuroendocrinol. 30, 46–94.
Ogawa, S., et al., 1998. Roles of estrogen receptor-alpha gene expression in reproduction-related behaviors in female mice. Endocrinology 139 (12), 5070–5081.
Parhar, I.S., et al., 1995. Embryonic development of gonadotropin-releasing hormone neurons in the sockeye salmon. J. Comp. Neurol. 362, 256–270.
Parhar, I.S., et al., 1996. Gonadotropin-releasing hormone gene expression in teleosts. Mol. Brain Res. 41, 216–227.
Pfaff, D.W., 2016. A Neuroscientist Looks at Robots. World Scientific Press, Singapore.
Pfaff, D.W., 2017. How the Vertebrate Brain Regulates Behavior, Direct from the Lab. Harvard University Press, Cambridge.
Pfaff, D.W., 2019. How Brain Arousal Mechanisms Work. Cambridge University Press, Cambridge.
Pfaff, D.W., Tabansky, I., Haubensak, W., 2019. Tinbergen's challenge for the neuroscience of behavior. Proc. Natl. Acad. Sci. U. S. A. 116 (20), 9704–9710.
Quiñones-Jenab, V., et al., 1997. Effects of estrogen on oxytocin receptor messenger ribonucleic acid expression in the uterus, pituitary and forebrain of the female rat. Neuroendocrinology 65, 9–17.

Ragnauth, A., et al., 2005. Female oxytocin gene knockout (OTKO) mice in a semi-natural environment (SNE) display exaggerated aggressive behaviors. Genes Brain Behav. 4 (4), 229–239.
Ribeiro, A.C., et al., 2012. siRNA silencing of estrogen receptor-α expression specifically in medial preoptic area neurons abolishes maternal care in female mice. Proc. Natl. Acad. Sci. 109 (40), 16324–16329.
Saltzman, W., Maestripieri, D., 2011. The neuroendocrinology of primate maternal behavior. Prog. Neuro-Psychopharmacol. Biol. Psychiatry 35 (5), 1192–1204.
Schwanzel-Fukuda, M., Pfaff, D.W., 1989. Origin of luteinizing hormone-releasing hormone neurons. Nature 338, 161–164.
Schwanzel-Fukuda, M., Bick, D., Pfaff, D.W., 1989. Luteinizing hormone-releasing hormone (LHRH)-expressing cells do not migrate normally in an inherited hypogonadal (Kallmann) syndrome. Brain Res. Mole. Brain Res. 6 (4), 311–326.
Sisk, C.L., 2016. Hormone-dependent adolescent organization of socio-sexual behaviors in mammals. Curr. Opin. Neurobiol. 38, 63–68.
Taroc, E.Z.M., et al., 2019. GnRH-1 neural migration from the nose to the brain is independent from Slit2, Robo3 and NELL2 Signaling. Front. Cell. Neurosci. 13, 70.
Tinbergen, N., 1963. On aims and methods of ethology. Z. Tierpsychol. 20 (4), 410–433.
Trainor, B.C., Sisk, C.L., Nelson, R.J., 2017. Hormones and the development and expression of aggressive behavior. In: Pfaff, et al. (Eds.), Hormones, Brain and Behavior, second ed. Academic Press (Elsevier), San Diego, CA, pp. 167–211.
Vasudevan, N., et al., 2001. Differential interactions of estrogen receptor and thyroid hormone receptor isoforms on the rat oxytocin receptor promoter leads to differences in transcriptional regulation. Neuroendocrinology 74 (5), 309–324.
Wu, Z., et al., 2014. Galanin neurons in the medial preoptic area govern parental behaviour. Nature 509 (7500), 325–330.

Further reading

Adolphs, R., Tusche, A., 2017. From faces to prosocial behavior: cues, tools, and mechanisms. Curr. Dir. Psychol. Sci. 26 (3), 282–287.
Archie, E.A., et al., 2014. Social affiliation matters: both same sex and opposite sex relationships predict survival in wild female baboons. Proc. R. Soc. Lond. Series B 281, 20141261.
Balasubramanian, R., Crowley Jr., W.F., 2017. Reproductive endocrine phenotypes relating to CHD7 mutations in humans. Am. J. Med. Genet. C Semin. Med. Genet. 175 (4), 507–515.
Chodorow, N., 1982. The Reproduction of Mothering. Harvard University Press, Cambridge.
Ciocchi, S., et al., 2010. Encoding of conditioned fear in central amygdala inhibitory circuits. Nature 468 (7321), 277–282.
Crowley, W.R., et al., 1985. From dud to stud: copulatory behavior elicited through conditioned arousal in sexually inactive male rats. Horm. Behav. 19, 227–230.
Fahrbach, S.E., Pfaff, D.W., 1986. Effect of preoptic region implants of dilute estradiol on the maternal behavior of ovariectomized nulliparous rats. Horm. Behav. 20, 354–363.
Fahrbach, S.E., Morrell, J.I., Pfaff, D.W., 1986. Effect of varying the duration of pre-test cage habituation on oxytocin induction of short-latency maternal behavior. Physiol. Behav. 37, 135–139.
Ganglberger, F., et al., 2018. Predicting functional neuroanatomical maps from fusing brain networks with genetic information. NeuroImage 170, 113–120.
Goodson, J.L., 2008. Nonapeptides and the evolutionary patterning of sociality. Prog. Brain Res. 170, 3–15.

LeDoux, J.E., 2000. Emotion circuits in the brain. Annu. Rev. Neurosci. 23, 155–184.
LeDoux, J.E., 2014. Coming to terms with fear. Proc. Natl. Acad. Sci. 111 (8), 2871–2878.
Numan, M., Morrell, J.I., Pfaff, D.W., 1985. Anatomical identification of neurons in selected brain regions associated with maternal behavior deficits induced by knife cuts of the lateral hypothalamus. J. Comp. Neurol. 237, 552–564.
Silk, J.B., et al., 2003. Social bonds of female baboons enhance infant survival. Science 302, 1231–1234.
Tovote, P., et al., 2018. The basolateral amygdala is essential for rapid escape: a human and rodent study. Cell 16 (1), 436–452.
Wilensky, A.E., et al., 2006. Rethinking the fear circuit: the central nucleus of the amygdala is required for the acquisition, consolidation, and expression of Pavlovian fear conditioning. J. Neurosci. 26 (48), 12387–12396.

CHAPTER 5

Visual add-ons during vertebrate evolution

When we see a friend and get a warm feeling, how does that work? What are the pathways from the eye to areas of the brain that register "friend" and then happiness? The serial, split-second steps are an important substrate of the social impulse which, in part, depends on facial recognition. To understand the process, we must first trace visual signals from the eye to the mid- and forebrain. For the neuroscientist, this is business as usual. Then, surprisingly, evidence suggests that nerve cells in the cerebral cortex participate actively in "top-down" control of visual perception, i.e., they may interpret the image: "Okay, that's my friend!" The last step—that of actually triggering the social impulse—is somewhat trickier, but there is exciting evidence from both human and non-human primate brains of neurons specialized for prosocial behaviors.

The smooth operation of this visual/interpretive brain mechanism suggests that the social impulse is *supposed* to occur. It's wired (by evolution) into structures that determine how we react to what we see. We don't decide whether to move towards another human being; our brains are designed to naturally move us in that direction. As with virtually all the phenomena that I discuss in this book—liking sex, responding to a baby—our prosocial movements in response to visual stimuli are virtually automatic. They evolved to be that way. The visual pathway associated with face recognition (and with recognition of the human form in general) should not be considered in isolation, e.g., as some sort of esthetic component of vision. Rather, it is intrinsic to a complex mechanism that, assuming it works as it should, results in our undertaking a social action. Visual perception can serve as the initial feature of a defined series of steps culminating in our involvement with other human beings.

In this chapter I do not try to argue about evolution step-by-step from the simplest animals to the human condition. Instead I highlight interesting features of human visual perception which were absent in the simplest animals. I assert that the only way these features could have shown up in

Origins of Human Socialization
https://doi.org/10.1016/B978-0-323-85861-8.00014-1

humans is by the process of evolution. So, once again, we see that evolution and brain mechanisms are tied together. Brain mechanisms exist *because* of evolution. Particular mechanisms (like visual acuity) are adaptations to the environment that, in humans, have enabled the species to survive. Survival is what drives evolution. Thus, the Mexican blind cavefish (*Astyanax mexicanus*) *lost* its eyes to evolution because, in the low-food/low-oxygen environment that they inhabited, vision actually consumed energy that was better spent in the onerous search for food. Of course, while the result seems amazing, in survival terms it makes sense. It tells us why, in terms of *Homo sapiens* and other species used by neurophysiologists, vision should be viewed (so to speak) from the perspective of species survival.

Pathways to vision

The easiest way to begin sketching out this process is to describe the classical neuronal pathways that serve visual perception. Neurons in the retina group their axons into the optic nerve, which carries their visual signals into other parts of the central nervous system. By far, the biggest pictures they deliver go to two places.

First, in primates, including humans, these pictures go to part of the thalamus, the dorsal lateral geniculate nucleus (Poggio, 1974). There, cells are ordered very neatly in layers. The distribution of optic nerve fibers in the thalamus is highly organized, faithfully representing the retinal surface in a so-called topographic fashion (meaning that the map of the retinal surface could be superimposed onto the map in the lateral geniculate nucleus). The central retina "has an extremely rich representation in the middle portion of the posterior two thirds of the lateral geniculate nucleus" (p. 499). Then, the axons of these highly spatially ordered lateral genicular neurons will radiate into the primary visual cortex. By this point in the visual system, the right primary visual cortex responds to stimuli in the left visual field, and vice versa. Below I will discuss the electrophysiological results in the cortex. It is sufficient to say here that from primary visual cortex a "ventral stream" of neuronal processing springs that helps detect curvatures and shapes of objects (Elder, 2018). Relative to that, a "dorsal stream" involves the visual control of action (Gallivan and Goodale, 2018).

The shortest possible summary would be to say that with respect to the object envisioned, the ventral stream determines "what" while the dorsal stream determines "where." It's more functionally accurate to say that the ventral stream is for perception (a cognitive process), and the dorsal stream is for directing action.

The other major axonal projection of the optic nerve is into the midbrain. There, a more primitive area of the brain, the left and right superior colliculus form two bumps on top of the midbrain (colliculus comes from the Latin for "little hill"). Clearly, the superior colliculus neurons serve different functions in visually guided behavior. They are essential "for the control of eye and head movements and more generally in visuomotor mechanisms" (p. 511), with the result of accurately guiding the eyes and head towards objects of interest. It thus makes sense that axons project to the lower brainstem and cervical spinal cord from some superior collicular neurons.

The electrophysiology of these pathways makes for an exciting story whose fundamentals are largely attributable to Steve Kuffler. When I took neurobiology courses at Harvard Medical School, I heard lots of lectures but always knew that the world-famous Kuffler was behind the entire program. A small, quiet, thoughtful man, he inspired at least one generation of neurophysiologists. My Marine Biological Labs roommate, Nick Spitzer, now a distinguished professor at University of California San Diego, said that Steve taught by example how to "adhere to the highest standards" in science. David Van Essen, who became head of a huge neurobiology program at Washington University, Saint Louis, was impressed by Steve's "persistent, penetrating but always friendly interrogation and his determination to pursue a line of questioning relentlessly until he was (intellectually) satisfied." John Nicholls, who wrote *From Neuron to Brain* with Steve, found Steve's "experiments to have a unique personal stamp, style, imagination, flair, skill and elegance." When Steve allowed me to take his Woods Hole course on neurophysiological recording techniques, his conversations with colleagues featured incredibly long pauses and silences which reflected his thoughtful style.

At a time when no one seemed to understand how the eye's retina worked, Kuffler clarified things. Taking up the idea of a "receptive field" in space, previously pursued by Rockefeller University's Nobelist H. Keffer Hartline, Kuffler (1953) pioneered the use of small natural visual stimuli and mapped their responsive zones onto the surface of the retina, i.e., an area of the retina which must be stimulated by illumination necessary to cause electrical discharge in a particular retinal ganglion cell or nerve fiber. Ah hah! Use natural visual stimuli; small spots of illumination of carefully regulated intensities. In my view, Kuffler's paper launched the entire field of mammalian visual system neurophysiology.

A few years later, at MIT, Jerry Lettvin et al. (1959) reported on "What the frog's eye tells the frog's brain." Visual inputs from, say, the image of a fly which the frog wanted to eat, were transmitted directly to the superior colliculus in the frog's midbrain and there directed the rapid tongue flick which caused the death of the fly and fed the frog.

The dynamic duo of visual neurophysiology were the Swedish doctor Torsten Wiesel and the Canadian David Hubel. Working together at Harvard Medical School, they moved mountains. As an MIT graduate student permitted to transfer into their courses, I was one of the few outsiders who actually witnessed one of their experiments. One of their later classics which, in the ensemble, led to their Nobel prize, was a 1968 paper which reported electrical recording from neurons in the visual cortex of non-human primates. They stated that "cells can be categorized as simple" (they possess distinct excitatory and inhibitory subdivisions—illumination of part or all of an excitatory region increased the maintained firing of the cell); or complex ("these respond to variously-shaped stationary or moving forms in a way that could not be predicted from maps made with small circular spots"), or hypercomplex. A small proportion of the cells were color coded.

Overall, their work showed that "the (visual) cortex is seen as a system organized vertically and horizontally in entirely different ways. In the vertical system (in which cells lying along a vertical line in the cortex have common features) stimulus dimensions such as retinal position, line orientation, ocular dominance, and perhaps directionality of movement, are mapped in sets of superimposed but independent mosaics. The horizontal system segregates cells in layers by hierarchical orders, the lowest orders (simple cells monocularly driven) located in and near layer IV, the higher orders in the upper and lower layers."

All of this work, from Kuffler and Lettvin to Hubel and Wiesel, laid the groundwork for the advanced neurophysiology which would come closer to answering how we appreciate other individuals—the objects of our social impulse—when we see them. We are not there yet, but we are making progress. So, the point that I want to make here is that understanding the intricate brain mechanism involved in identifying another person is crucial to why brain mechanisms are a prong of my theory explaining human sociality. These mechanisms go to the heart of sociality—if you cannot identify a shape as a person; a face as belonging to a specific person; and the location of that person, then you cannot move in that person's direction. In effect, you cannot be social if the object of sociality is not discernable. Our visual/interpretive equipment, from eye to brain, leads to sociality.

Active seeing/viewing

For many years, I had understood the visual system in terms of its being a passive recipient and interpreter of visible light signals. Sure, this could be complex. My department head at MIT, Hans-Lukas Teuber would quip "There's more to vision than meets the eye." His early work had shown the complexities of visual perceptual losses following penetrating brain injuries suffered during World War II. Nevertheless, all of that early work tended to show that the flow of information was one-way, "from the retina on up." So, when my friend and Rockefeller colleague, professor Charles Gilbert, began to talk about top-down controls in the mechanisms for visual attention and perception, I initially had a skeptical response. Top-down (illustrated below) means that brain regions beyond those traditionally regarded as visual sensory pathways can influence activity in those sensory pathways (Fig. 1).

But Charles was right. It turns out that a large and growing body of electrophysiological and behavioral data, coupled with clever experimental designs and quantitative thinking, have not only proven that that behavioral context can influence the responses of visual cortical neurons, but these data also begin to tell us exactly how that occurs.

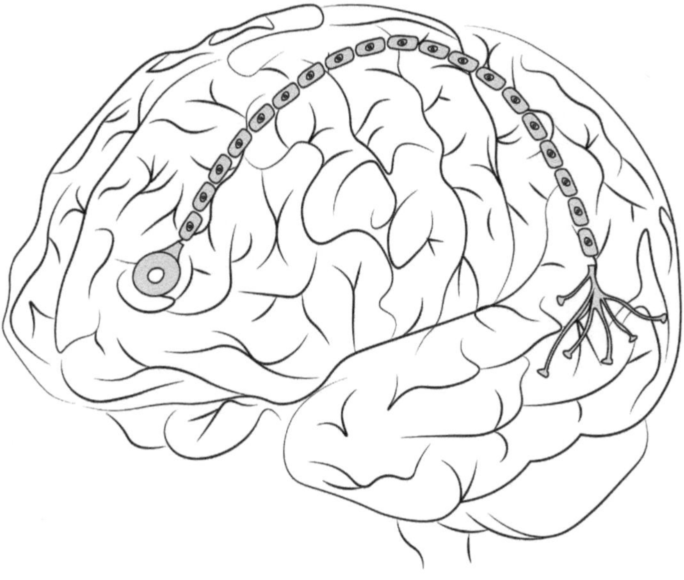

Fig. 1 Neurons in the prefrontal cortex (left) of experimental animals can influence neurons in the visual cortex (right). This fact offers an explanation of "top down" influences on vision, according, for example, to expectation or motivation.

For me, the top-down story begins with a quest for the explanation of visuospatial attention (Gilbert et al., 2000). Such a mechanism would be crucial for drawing the visual system into the world of social instincts. For example, the vision of an attractive woman across the room might quickly draw a man's attention. Following on from Gilbert's work, MIT's Robert Desimone, with Fries (2008), emphasized synchronization of firing among neurons activated by an attended stimulus. They suspect that "precise synchronization could enhance the postsynaptic impact of cells carrying behaviorally relevant information." Not only was there an enhancement of synchronization during visual stimulation, but also "selective attention significantly changed the prestimulus pattern of synchronization." The idea of the importance of temporally grouped neuronal activity is supported by "simultaneous recordings from populations of individual cells which have shown that attention is associated with changes in the correlated firing of neurons that can enhance the quality of sensory representations" (Maunsell, 2015).

Most interesting, Desimone and Gregoriou (Gregoriou et al., 2012) linked shifts of attention to cortical neurons responsible for eye movement, themselves a "source of top-down feedback" to high level visual cortex.

Gilbert's thoughts about top-down regulation were confirmed and extended by Buffalo and Desimone (2010), who compared the magnitude and latency of attentional enhancement of firing rates in V1, V2, and V4 in the same animals performing the same task. Buffalo and *Desimone* found a reverse order of attentional effects, such that attentional enhancement was. They found that attentional effects on stimulus processing were larger and earlier in higher order visual cortex than in primary visual cortex. Further, in a landmark paper in Science, Gregoriou and Desimone (2009) used electrical recordings in humans and monkeys to show attentional enhancement of visual cortex neuronal responses, and suggested this regulation is initiated by neuronal groups that regulate eye movement as well as nerve cells in prefrontal cortex (Gregoriou, 2009).

Historical perspective. Over decades, visual neurophysiology was conceived as a description of the receptive fields and stimulus properties required for a thalamic or cortical neuron to fire a train of action potentials. But such discoveries by themselves would not explain visual attention. It turns out that the first steps of explanation dated back to Kuffler's description of a "non-classical" receptive field. Kuffler emphasized that "… not only the areas from which responses can actually be set up by retinal illumination may be included in a definition of the receptive field but also all areas which show a functional connection, by an inhibitory or excitatory effect on a ganglion cell. This may well involve areas which are somewhat remote from a ganglion cell and by themselves do not set up discharges" (Kuffler, 1953).

Now my historical view jumps from the retina to the cortex in order to explain visuospatial attention, Gilbert cited long-range connections within visual cortex whose activity could be shaped by higher order cognitive processes including motivation and learning. As a consequence, we should view a visual cortex neuron as "a dynamic and active processor, subject to top-down influences."

How should we study contextual influences on a visual cortex neuron's response? Kapadia et al. (2000) provides an important example. Depending on the arrangement of visible lines definitely outside the neuron's strictly defined receptive field, the train of action potentials could be modulated in a non-linear way. For one arrangement of lines, "the response grew to more than three times the baseline level" even though those extra lines by themselves elicited little response. When a different arrangement of those extra lines was presented, they had the opposite effect. They inhibited the unit's response to the central stimulus below the baseline level. That is, for any given visual-signaling neuron in the cerebral cortex, the effects of context (as would require "top down" influences) are not constant across the receptive field of that neuron. Charles Gilbert's approach transformed that phenomenon of extra receptive-field responses from a simple annoyance to a principle finding that underlies modern cortical physiology.

In this tradition, the results of Chen et al. (2008) were impressive. For their visual cortical neurons, "task difficulty modulated neuronal firing rate at the earliest stages of cortical visual processing; and increasing task difficulty enhanced V1 neuronal firing rate at the focus of attention and suppressed it in regions surrounding the focus." Likewise, for Motter (1993), visual cortical neurons displayed differential sensitivity when *attention* was directed towards versus away from the spatial location of the receptive field just before and during stimulus presentation. That is, representation of the neural activity especially in downstream visual cortex highlights the potential targets in the visual scene at the expense of background objects (Motter, 1994). Treue and Martinez Trujillo (1999) followed up demonstrations of changes in neural responses based on spatial attention by demonstrating *non*-spatial, feature-based attentional modulation of visual motion processing: attention increased the gain of direction-selective neurons in a specific visual cortical area without narrowing the direction-tuning curves. Finally, before Robert Desimone moved from the NIH to MIT, his team (Chelazzi et al., 1999) theorized that a "biased competition" model of attention would account for recoding data, an idea in which objects in the visual field compete for representation in the cortex, and "this competition is biased in favor of the behaviorally relevant object by virtue of *top-down* feedback from structures involved in working memory."

For purposes of this book, the ability to pay attention to a real or potential social partner is crucial. If you cannot see and identify someone, and determine where they are situated, then social approach is impossible. Social approach is *always* initiated by sensing another person. First comes the sense impression, then comes the social impulse. For most of us, our visual capacity is integral to the social impulse. It establishes that a person, even a specific person, is within a visual field. It enables us to move in their direction. Our visual capacity is not just a "brain mechanism" because the brain interprets data from the eye; it is a brain mechanism because that data results in further signals that result in physical movement. We take some sort of action directly traceable to images registered on brain cells. It is because of such responses that I identify "brain mechanisms" as one factor (among three related factors) that account for the initiation of human sociality.

Gilbert and Li (2013) attacked the problems of visual attention and visual learning by taking steps beyond the usual formulation of visual neurons as slaves to their inputs from the visual thalamus. Instead, they viewed visual cortex neurons as "adaptive processors that are able to assume different functional states according to the task being executed." As part of this formulation, it is possible to say that visual cortex neuronal systems can "multiplex"—individual neurons can perform more than one type of function, perhaps even at the same time. Different signaling requirements can run through the same neuron or, perhaps, the same visual cortical column. Importantly, Gilbert and his colleagues recorded task-dependent changes in neural tuning and information content, even in primary visual cortex, which would require modulation of visually receptive cortical neurons by inputs from elsewhere in the cortex. "Top-down influences dynamically change effective connectivity within and between cortical areas, allowing neurons to take on functional properties, that are appropriate for the immediate behavioral context." Thus, for example, attention to a loved one or motivated valuation of a person's image, could be explained by lateral-running signals coming, notably, from the prefrontal cortex.

University of Chicago professor John Maunsell's work (Luo and Maunsell, 2018) could support the view that some aspects of visual perception, for example in social situations, would require participation of neurons outside visual cortex. Consider that visual performance in such situations involve multiple components of attention. Luo and Maunsell recorded from monkey lateral prefrontal cortex and found that, in contrast to visual cortex, neurons in prefrontal cortex changed their firing rates and interneuronal correlations when subjects changed either their behavioral performance criterion or their sensitivity. "These results indicate that attention-related neuronal modulations in separate brain regions" can be related to visually guided behavior (as in society) in a manner that primary visual cortical neurons are not. Likewise, neurons in higher order visual cortex, called V4, could show "attention-related neuronal modulations corresponding to behavioral shifts" in ways that primary visual cortical neurons do not (Luo and Maunsell, 2015). Likely, some of those modulations were due to signals from prefrontal cortex. The most complicated aspect of this story involves the idea of visual response "normalization." The term "normalization" means different things in different areas of science. Maunsell means "a computation by which neuronal responses are modified by stimulus context." Recording from cortical neurons outside visual cortex, Maunsell reported what he termed a spatially tuned sensory normalization mechanism (Ni and Maunsell, 2019). Normalization as used here describes a widely used calculation that results in a response function ranging between zero and one and comprises what has been described as a "tuning function scaled by the contrast of the stimulus." The bottom line is that—when any of us fixes our visual attention on a friend or enemy—the degree of computational sophistication which goes into determining our response is very high.

In other words, we are designed to fix another person within our field of vision. We are not designed to reach out randomly—blindly, so to speak—but to move towards people whom we know, who are identifiable as enemies or fiends, i.e., who are in our intellectual and emotional field of vision. Our literal field of vision is designed to work in tandem with whom and what we know already. When America's battles were fought on the ground in close combat, e.g., during the Revolutionary and Civil Wars, soldiers wore uniforms that were instantly distinguishable so that everyone knew whether to offer support in one direction or—in the other—rush their enemy with a bayonet. The same principle applies to football teams. Human sociality would be the palest image of itself were it not for our acute vision, and our ability to interpret almost instantly what we see.

Needless to say, we have now contrived to augment our vision exponentially, so that we can see objects too small or too distant to be visible to the naked eye. We have, again, augmented evolution with technology. Arguably, microscopes and telescopes promote our survival—with the one, we can see harmful microorganisms, and so defeat them; with the other… well, there are sunspots and asteroids that could disrupt our lives, and maybe someday we'll *really* need to emigrate. But the point is that vision is continuing to evolve beyond its initial limitations. We are able to think visually in ways that would have been inconceivable before our primate visual capabilities evolved. We are enhancing our survival capabilities with technology.

A very sensible pattern of results regarding the shape of visual stimuli came from Warren Pettine in Tirin Moore's lab at Stanford. Pettine et al. (2004) showed evidence of a division of labor between neurons recorded within the superficial and deep layers of visual cortex in sensory coding and behavior. Neurons within the superficial layers of extrastriate area V4 outperformed their deep layer counterparts in coding the shape of visual stimuli, both in latency and accuracy. In contrast, neurons within the deep layers outperformed their superficial layer counterparts in the coding of selective attention and the planning of eye movements, particularly the latter. Thus, neurons superficial, out near the borders of the cortex, are stimulus driven. Neurons deep in the cortex are behavior correlated. What Pettine's pattern of results means is that in cortical neuronal columns, as signaling goes from superficial to deep layers, the information is processed in a manner that depends on behaviorally relevant signals coming in from other cortical regions. This is an especially sensible application of top-down controlling.

Some work concerning top-down visual signaling control uses the idea of predictive coding. Choi et al. (2018) used the trick of partial visual occlusion, i.e., hiding, to test the ability of the primate visual system to discriminate shapes. Their results suggested that inference of what shape is actually there, occluded or not, occurs in two steps. "Initially, V4 responses are driven solely by bottom-up sensory input and are thus strongly influenced by the level of occlusion. After a delay, V4 responses combine both feedforward input and feedback signals from the PFC; the latter reflect predictions made by PFC about the visual stimulus underlying V4 activity." Thus, according to Choi et al., higher level visual cortex neurons participate in hierarchical inference, with feedback signals encoding top-down predictions about occluded shapes.

Li et al. (2004) published neurophysiological results which demand explanation by top-down influences. Visual cortical neurons took on novel functional response properties related to the attributes of behavioral training to which the animals were exposed. Further, these response properties depended on the perceptual task being performed. As one example, when the animal performed a visual learning task where side flanking lines were task-relevant, some visual cortical nerve cells showed substantial positive modulation by the appearance of these side flanking lines. In contrast, when the animal performed the task where the side flanks were task-irrelevant, there was weak negative modulation for the same nerve cells. This kind of learning-dependent modulation demands participation by other cortical areas in the coding.

I should add that not all top-down results and theories are limited to the cortex. In Munich, psychology professor Karl Finke and his colleagues (Menegaux et al., 2019) came up with imaging evidence for "recurrent loops between the posterior thalamus and posterior visual cortices to be relevant for generating attentional weights for competing objects and for maintaining selected objects" in visual memory. Thalamic structures and their connections to visual cortex may be important for top-down control and visual memory.

Professors George Reeke and Charles Gilbert at Rockefeller (Piëch et al., 2013) used top-down thinking and computer modeling to explain the grouping of oriented line segments that define object boundaries in complex visual scenes. That is, they wanted to answer the question of how we know what belongs to an object visualized and what is part of the background. They achieved a mathematical model which "displays contour-related facilitation of neuronal responses." Their quantitative computer model matches the results of neuronal electrical recording experiments in visual cortex, and even reproduced the results of visual perceptual learning. Thus, the next time you recognize the visual image of your mate, your cortex might be applying the math conceived in this paper.

In summary, neurobiologists are beginning to understand in increasing detail how we interpret the shape in our field of vision as the image of a friend. As Gilbert states: "cortical neurons are subject to top-down influences of attention, expectation and perceptual tasks. 'Top-down' refers to cognitive influences and higher-order representations that impinge on earlier steps in information processing." Further, understanding how lines and curves in an image support the formation of perception of objects—in our case social objects—brings us closer to explaining brain mechanisms for visual system participation in our social lives. In effect, the social impulse begins with information processing. The speed with which we act on that impulse is owing, in large part, to the speed with which the brain processes visual information, allowing us not just to take action but to take appropriate action. That is, though perception of another will result in an initial movement *towards* that other, such perception will result in a particular initial movement *calibrated* to the image of whom we actually perceive. How we move towards a friend or a newborn is vastly different from how we move towards an attractive member of the opposite sex or a dignitary in a receiving line at a garden party.

In each of those cases, we move. But the complexity of how we are able to see and interpret shapes has its appropriate result in the widely varied actions that we take as a consequence. Thus, while I am primarily concerned here with only the initial social spark, i.e., the very first movement towards another human, that movement will take any number of shapes based on the shape and identity of whom we see, in the setting that we see them.

Seeing social—Face recognition

The purpose of this chapter is to highlight mechanisms for visual perception in a social setting. When we see a friend, what actually happens so that we recognize that person? Obviously, the issue is crucial because if we do not know a person—if we cannot identify them in some way or if we do not see them as attractive—then no movement in their direction will occur. If we think of life as one vast Brownian movement of people flashing in an out of our field of vision over the course of our lives, we only move towards a tiny fraction of them. Why? What makes us discriminate between the few whom we do approach and all the billions of others in trillions of potential encounters? It gives one pause just to imagine the volume of work imposed on the brain as it continually toggles between Yes and No or, rather, Go Ahead and Don't Bother.

Much of this movement depends on face recognition. "Oh, I know that person" or "That person *looks* interesting." Face recognition (like voice recognition) technology is becoming crucial in law enforcement, and its practitioners have still not entirely mastered it (e.g., there are claims that machines fail to recognize black faces because of bias in the inputted data). Humans are better at face recognition than machines: not only can we identify particular faces, but we can tell whether someone is happy, sad, or perplexed. We can recognize emotions, for example, in a smile or a frown. In other words, our brain skillfully interprets when and how we should move towards someone, i.e., when and how to activate our social impulse, based on its recognition (or non-recognition) of faces. Face recognition has evolved as a component of the social impulse.

When I was a graduate student, we joked about a "grandmother cell" in the visual cortex—a *reductio ad absurdum* reference to an imagined neuron that fired action potentials only if the image of our grandmother's face was presented to the retina. We were sure that visual coding would be more complicated. But we were wrong.

Charlie Gross was a postdoctoral fellow at MIT at the same time that I was and, according to Earl Miller and Robert Desimone, our department head Hans-Lukas Teuber "encouraged Charlie and his colleagues to use a single neuron electrode approach to study the inferior temporal cortex in the monkey, which was virtually uncharted territory but was thought to possibly be a repository of memories." Charlie knew enough about inferior temporal cortex to predict that "neurons would be selectively activated by complex stimuli." Years later, as a young professor at Princeton, Charlie discovered neurons in this part of cortex which are selectively activated by the sight of faces.

Later, Robert Desimone (Desimone et al., 1984) was a graduate student studying the inferior temporal cortex with Charlie, recording from individual cortical neurons. He stated that "most such neurons do not appear to be detectors for complex objects, but found a population of cells that responded selectively to faces. The responses of these cells were dependent on the configuration of specific face features, and their selectivity was maintained over changes in stimulus size and position."

These groundbreaking discoveries demanded to be followed up, and MIT professor Nancy Kanwisher (1997) did that. Using functional magnetic resonance imaging (fMRI) with human subjects, she "found a specific cortical area in the back of the human brain called the fusiform gyrus that was significantly more active when the subjects viewed faces than when they viewed assorted common objects."

Her discoveries over the years speak to the issues of visual attention, in this case to human faces, addressed in Charles Gilbert's experiments, cited above. In later years Kanwisher reported that this fusiform gyrus area also responded to voices, thus offering the possibility of a cross-sensory-modality integrated response to the presence of another person.

In her 2006 review, Kanwisher emphasizes not only the importance of face recognition in our social lives, but also highlights the ability of the human cerebral cortex "to extract this information within a fraction of a second of viewing a face, important for normal social interactions.... an ability which... has probably played a critical role in the survival of our primate ancestors." She is talking about specificity for the recognition of human faces, as separate from other objects, animals (Kanwisher et al., 1999), and faces presented upside-down (Kanwisher et al., 1998).

This makes sense. After all, in normal social life, how often do we look at people upside-down? New questions (see below) which follow up functional magnetic resonance imaging will be answered by electrical recordings from individual cortical neurons.

Indeed, a long series of studies, beginning with cooperation between Winrich Freiwald and Doris Tsao (2003) at MIT and Harvard, pushed forward our understanding of the brain's face recognition system. They used fMRI techniques to reveal "face-specific patches in the left hemisphere (in the left posterior inferior temporal gyrus and in the fusiform gyrus) and two face- specific patches in the right hemisphere (in the anterior superior temporal sulcus and in the fusiform gyrus." Their discoveries further reinforced the marked impact the need for face recognition has on the design of visual perception pathways in the cortex.

Technical background to these discoveries: A face-specific "patch" is defined as part of an image of the brain in magnetic resonance imaging, a technique which depends on the ability of certain atomic nuclei to absorb or emit energy when placed in a strong magnetic field. As reviewed by Bandettini (2014), "it is the extreme sensitivity to subtle changes in blood susceptibility arising from changes in blood oxygenation following localized activation-induced flow changes that allows blood oxygen level-dependent (BOLD) fMRI to be a viable functional contrast. The head, and therefore the brain, is placed in a device which produces a strong magnetic field which is oscillating at a certain well-chosen frequency. With this sensitivity, one can extract substantial neuronal and physiologic information. The fMRI signal changes—both the magnitude and dynamics—are influenced by the interplay of changes in cerebral metabolic rate, blood flow, and blood volume. The hemodynamic response is a nonlinear function of neuronal activity at very brief neuronal activity durations or at very low neuronal activity magnitudes" (reviewed, Bandettini, 2014). Functional MRI is non-invasive and can yield pictures in all three planes of brain sections: anterior/posterior, dorsal/ventral and medial/lateral. Rapid sequences of images spread over milliseconds allow the investigator to track responses to specific stimuli. The functional MRI signal does not reflect the activity of an individual neuron, but must comprise a group of neurons which, when activated, require oxygenated blood which supports their use of metabolic energy to produce electrical action potentials.

With functional MRI, Freiwald and Tsao found five face-specific patches in the monkey brain which "were statistically robust and reliably imaged across ten experimental sessions spanning almost nine months."

How can these exciting results be interpreted in terms of the electrical activity of individual neurons? The use of small, highly selective electrodes lowered into the monkey cerebral cortex and positioned with a few microns of neurons in face patches did the job (Tsao and Freiwald, Science 2006). The average ratio of excitation of face to non-face stimuli was very high, with a ratio of face to non-face as high as 21. Further, Moeller, Freiwald and Tsao (2008) used "electrical microstimulation combined with simultaneous functional magnetic resonance imaging. Stimulation of each of four targeted face patches produced strong activation, specifically within a subset of the other face patches." Thus, in face recognition we deal with a network of face-responsive cortical neurons.

Freiwald et al. (2009) followed up. They "found that middle face patch neurons detect and differentiate faces using a strategy that is both part-based and holistic. Cells detected distinct constellations of face parts. Furthermore, cells were tuned to the geometry of facial features." Further, Landi and Freiwald (2017) discovered locations in the monkey cortex outside the core face-recognizing network, for which the sight of familiar faces led to significantly larger electrophysiological (fMRI) responses. The familiar face stimuli could lead to a sudden electrophysiological response, in time, reminiscent of the sudden recognition of a familiar person which we sometimes experience.

Michael Platt, at Duke University, was thinking along similar lines to other neuroscientists (Klein et al., 2009). Aware that humans pay attention to members of their groups, he states, "This process likely proceeds via two distinct but integrated pathways: an ancestral, subcortical route that mediates crude but fast orienting to animate objects and faces; and a more derived route involving cortical orienting circuits that mediate nuanced and context-dependent social attention." Even primitive primates (lemurs) demonstrate Platt's point. Our results support the hypothesis that stem primates were capable of orienting towards and following the attention of other individuals (Shepherd and Platt, 2008). In fact, amazingly, monkeys can show "preferences for brand logos (associated with sex and status) even though choosing them provided no tangible rewards." The authors speculate that "the power of sex and status in advertising emerges from the spontaneous engagement of shared, ancestral neural circuits that prioritize information useful for navigating the social environment" (Acikalin et al., 2018).

In Michael Platt's research, social attention is subject to genetic and surgical manipulation. When Watson et al. (2015) tested monkeys with abnormally long alleles of the tryptophan hydroxylase gene responsible for manufacturing the neurotransmitter serotonin, those monkeys showed significantly less social vigilance than in Platt's behavioral assays. Further, Roy et al. (2014) studied gaze-following by monkeys, by which a monkey might discern another monkey's interests and intentions. Inactivating neurons in a "face region"—the posterior superior temporal sulcus—by microinjections of an inhibitory neurotransmitter, GABA, receptor agonist strongly suppressed gaze following. Thus, neurons in face regions are not just neurophysiological curiosities. They matter for visually guided behavior.

How long might a neuron's face recognition capacity last? McMahon et al. used chronically implanted micro-electrodes to show that "the (monkey) inferotemporal cortex contains face-selective cells that show virtually no change in their patterns of visual responses over time periods as long as one year" (McMahon et al., 2014). The authors inferred that such neurons "are specialized for aspects of face perception that demand stability" as opposed to flexible changes according to mood and experience.

These patches are connected with each other (Grimaldi et al., 2016). Doris Tsao's lab microinjected retrograde tracers into four different face patches. These tracers will be absorbed by synapses, travel back up the axons and reveal the nerve cell bodies which sent their axons to the microinjected brain region. They observed, "the patches are strongly and specifically connected to each other." We are not talking about a one-way street. "Patches were densely interconnected to one another in both feedforward and feedback directions."

While these face patches "act as distinct functional units rather than an integrated feedforward-processing pipeline" (Shepherd and Freiwald, 2018), they "exhibit some functional specialization." For example, neuronal groups could be selectively activated during the view the authors called "second person" (directly gazing at the subject) compared to "third person," (looking at an image of averted gaze).

Do these neurons with such selectivity for faces actually support face-recognition?

Sadagopan et al. (2017) used microinjections of an inhibitory transmitter (GABA) receptor agonist to block the electrical activity of face-selective neurons and measured performance in a naturalistic face-detection behavioral task. Such inactivation impaired face detection. Inactivation of other regions did not have that effect. Conversely, visual recognition of body and objects was not affected.

If the story of face recognition by neurons in the brain had stopped there, average readers might have been left with a feeling of irresolution. They might have asked what role this neuronal system plays in our daily lives. Investigators have sprung to the task and, over the years, answered in at least three ways. They have shown the influence of social values; the role of social status; and (most amazingly) the brain's perception of social interactions.

First, consider the influences of frontal cortical activity on the visual system, discussed above: those influences must convey evaluations of social targets for potential visual attention. That is what the prefrontal cortex does. Volkow et al. (2019) say that "reinforcing effects of drugs mostly depend on dopamine signaling in the nucleus accumbens, and chronic drug exposure triggers glutamatergic-mediated neuroadaptations predominantly in prefrontal cortical regions"; and further that "impaired prefrontal top-down self-regulation that favors compulsive drug-taking against the backdrop of negative emotionality." Dixon et al. (2017) state that neurons in portions of the prefrontal cortex *assign value* to particular sensory inputs, which would include social inputs. Moreover, this is a two-way street. Social play influences the dendritic complexity of nerve cells in the medial prefrontal cortex (Himmler et al., 2018).

Second, Koski et al. (2017) worked with human subjects and tested the hypothesis that social status information is embedded in our neural representations of other individuals. These participants had "learned to associate faces with names, job titles that varied in associated status, and explicit markers of reputational status." Then, concentrating on the lateral orbitofrontal cortex—"a region involved in social and nonsocial valuation," they could decode imaging results faces and, startlingly, identify occupations based on their status. Their data suggested that a "face-selective region of the lateral orbitofrontal cortex may contribute to the perception of social status, potentially underlying the preferential attention and favorable biases humans display toward high-status individuals."

With similar intentions, Munuera et al. (2018) studied monkey brain and behavior to test "The social brain hypothesis which posits that dedicated neural systems process social information." They showed "that the primate amygdala encodes the hierarchical rank of individuals in the same neuronal ensembles that encode the rewards associated with nonsocial stimuli." At least for the amygdala, they inferred that "information about hierarchical rank-which contributes to the assessment of the social value of individuals within a group-is linked in the amygdala to representations of rewards associated with nonsocial stimuli."

Then came the startling discovery of Sliwa and Freiwald (2017). As mentioned, they discovered a network centered in the medial and ventrolateral prefrontal cortex that is exclusively engaged in social interaction analysis (Fig. 1 in Chapter 4). The exclusivity of this type of specialization for social interaction was found for nowhere else in the brain. "Large parts of the social interaction network were exclusively selective for social interactions and did not respond to any other stimulus condition in the context of the present design." Areas of this exclusively social interaction network centered upon clusters in the prefrontal cortex. That is, a "statistical map of enhanced activation for the social interaction centered on the medial and dorsomedial prefrontal cortices."

Overview

In summary, Freiwald et al. (2016) have taken a "36,000-ft view" and can claim several victories for modern neurophysiology. Face detection, holistic facial feature processing, and responses to faces as sources of social information have all been dealt with successfully. All of this is incredibly interesting to techies and law enforcement authorities, whom I will discuss in the final chapter.

In *Human Ethology* (1989), Irenaus Eibl-Eibesfeldt affirms the significant roles played by visual signals in human interpersonal communication (p. 437). "Expressive movements are behavior patterns that have undergone distinctive differentiation in the service of signaling… (for example) women use tongue flicking at a distance to indicate contact readiness. But most frequently it is a completely unconscious expression of female heterosexual readiness." In contrast, "sticking out the tongue as a gesture of rejection is derived from refusal to take food" (p. 438). "The ritualization of behavioral patterns into signals" involves, for example, the simplification and rhythmic repetition and the amplification of the movement (p. 439). Eibl-Eibesfeldt emphasizes the cross-cultural validity of some of these social signals, key to guiding social approaches.

The fact is that across human societies, humans signal and move towards others in their field of vision. But ritualization of these movements, after the initial movement, only involves specific cultural adaptations of the universal instinct to move forward—to behave prosocially. As I have said repeatedly in this book, we should not confuse the product of intellection/thought with what is instantaneous, unreflective, and essentially a motor response to stimuli. The visual apparatus discussed in this chapter pertains emphatically to the latter—to instant responsiveness—and it is the same irrespective of cultural adaptations of our later movements after the initial approach is undertaken.

This chapter demonstrates that what we see and how we interpret it, including how we interpret faces, is a major brain mechanism that promotes prosocial behavior. You (and all of us) were designed to see faces and to make social contact. This is why, our visual apparatus should be considered part of the brain mechanism that makes responses appropriate to the social impulse possible. Vision also makes that impulse one that is consummately refined, since, just because someone is "there" does not automatically trigger approach. We move when it is appropriate, for example, towards someone whom we can identify or to whom we feel attracted. Otherwise, we would be moving in every direction at once, turned-around and over-excited. From this perspective, vision is as much a limiting factor as it is one that opens us up to reality. It orients us in the right direction to promote our survival and the survival of our species.

Our recognition of faces and social interactions comprise, of course, an element of an element of my tri-partite theory of how the social impulse originated. These capacities belong, generally speaking, to "brain mechanisms," which are connected as well to evolution and to babies' development. We cannot consider vision as a social brain mechanism *except* in connection with these other elements of my theory. Taken together, these elements have permitted *Homo sapiens* to survive. In fact, the importance of vision (among all our brain mechanisms allowing us to sense the world) cannot be overstated. We perceive up to 80% of our impressions by means of sight. In the English-speaking world, most people can describe visual and aural impressions more easily than they can describe tastes, or smells, or what they touch.

In the next chapter, I will discuss our interesting auditory capacities, which also can operate to orient us towards appropriate social contacts.

New questions

1. How will we get to understand, in detail, how "top down" regulation of visual signaling works to support social recognition and approach?
2. Facial recognition technology has gotten so good that it has become a threat to privacy. As it improves further, what will be the optimal balance among security and privacy?

References

Acikalin, M.Y., et al., 2018. Rhesus macaques form preferences for brand logos through sex and social status based advertising. PLoS One 13 (2), e0193055.
Bandettini, P., 2014. Neuronal or hemodynamic? grappling with the functional MRI signal. Brain Connect. 4 (7), 487–498.
Buffalo, E.A., et al., 2010. A backward progression of attentional effects in the ventral stream. Proc. Natl. Acad. Sci. U. S. A. 107, 361–365.
Chen, Y., et al., 2008. Task difficulty modulates the activity of specific neuronal populations in primary visual cortex. Nat. Neurosci. 11, 974–982.
Choi, H., Pasupathy, A., Shea-Brown, E., 2018. Predictive coding in area V4: dynamic shape discrimination under partial occlusion. Neural Comput. 30 (5), 1209–1257.
Desimone, R., et al., 1984. Stimulus-selective properties of inferior temporal neurons in the macaque. J. Neurosci. 4 (8), 2051–2062.
Dixon, M., et al., 2017. Emotion and the prefrontal cortex: an integrative review. Psychol. Bull. 143, 1033–1081.
Eibl-Eibesfeldt, I., 1989. Human Ethology. de Gruyter, New York.
Elder, J.H., 2018. Shape from contour: computation and representation. Annu. Rev. Vis. Sci. 4, 423–450.
Freiwald, W.A., Tsao, D.Y., Livingstone, M.S., 2009. A face feature space in the macaque temporal lobe. Nat. Neurosci. 12 (9), 1187–1196.
Freiwald, W., et al., 2016. Face processing systems: from neurons to real-world social perception. Annu. Rev. Neurosci. 39, 325–346.
Fries, P., et al., 2008. The effects of visual stimulation and selective visual attention on rhythmic neuronal synchronization in macaque area V4. J. Neurosci. 28 (18), 4823–4835.
Gallivan, J.P., Goodale, M.A., 2018. The dorsal "action" pathway. Handb. Clin. Neurol. 151, 449–466.
Gilbert, C.D., Li, W., 2013. Top-down influences on visual processing. Nat. Rev. Neurosci. 14 (5), 350–363.
Gilbert, C., et al., 2000. Interactions between attention, context and learning in primary visual cortex. Vis. Res. 40 (10–12), 1217–1226.
Gregoriou, G.G., 2009. Long-range neural coupling through synchronization with attention. Prog. Brain Res. 176, 35–45.
Gregoriou, G.G., et al., 2009. High frequency, long-range coupling between prefrontal and visual cortex during attention. Science 324, 1207–1210.
Gregoriou, G.G., Gotts, S.J., Desimone, R., 2012. Cell-type-specific synchronization of neural activity in FEF with V4 during attention. Neuron 73 (3), 581–594.
Grimaldi, P., Saleem, K.S., Tsao, D., 2016. Anatomical connections of the functionally defined "face patches" in the macaque monkey. Neuron 90 (6), 1325–1342.
Himmler, B.T., et al., 2018. Juvenile social experience and differential age-related changes in the dendritic morphologies of subareas of the prefrontal cortex in rats. Synapse 72 (4). https://doi.org/10.1002/syn.22022 (Epub 2017 Dec 21).
Kanwisher, N., McDermott, J., Chun, M.M., 1997. The fusiform face area: a module in human extrastriate cortex specialized for face perception. J. Neurosci. 17 (11), 4302–4311.
Kanwisher, N., Tong, F., Nakayama, K., 1998. The effect of face inversion on the human fusiform face area. Cognition 68 (1), B1–11.
Kanwisher, N., Stanley, D., Harris, A., 1999. The fusiform face area is selective for faces not animals. Neuroreport 10 (1), 183–187.
Kapadia, M.K., et al., 2000. Spatial distribution of contextual interactions in primary visual cortex and in visual perception J. Neurophysiology 84 (4), 2048–2062.
Klein, J.T., Shepherd, S.V., Platt, M.L., 2009. Social attention and the brain. Curr. Biol. 19 (20), R958–R962.

Koski, J.E., Collins, J.A., Olson, I.R., 2017. The neural representation of social status in the extended face-processing network. Eur. J. Neurosci. 46 (12), 2795–2806.
Kuffler, S.W., 1953. Discharge patterns and functional organization of mammalian retina. J. Neurophysiol. 16 (1), 37–68.
Landi, S.M., Freiwald, W.A., 2017. Two areas for familiar face recognition in the primate brain. Science 357, 591–595.
Lettvin, J., et al., 1959. What the frog's eye tells the frog's brain. In: Proceedings, Soc. of Radio Engineering, pp. 1–23.
Li, W., Piëch, V., Gilbert, C.D., 2004. Perceptual learning and top-down influences in primary visual cortex. Nat. Neurosci. 7 (6), 651–657.
Luo, T.Z., Maunsell, J.H., 2015. Neuronal modulations in visual cortex are associated with only one of multiple components of attention. Neuron 86 (5), 1182–1188.
Luo, T.Z., Maunsell, J.H.R., 2018. Attentional changes in either criterion or sensitivity are associated with robust modulations in lateral prefrontal cortex. Neuron 97 (6), 1382–1393.
Maunsell, J.H.R., 2015. Neuronal mechanisms of visual attention. Annu. Rev. Vis. Sci. 1, 373–391.
McMahon, D.B., Jones, A.P., Bondar, I.V., Leopold, D.A., 2014. Face-selective neurons maintain consistent visual responses across months. Proc. Natl. Acad. Sci. U. S. A. 111 (22), 8251–8256.
Menegaux, A., et al., 2019. Theory of visual attention thalamic model for visual short-term memory capacity and top-downcontrol: evidence from a thalamo-cortical structural connectivity analysis. Neuroimage 195, 67–77.
Moeller, S., Freiwald, W.A., Tsao, D.Y., 2008. Patches with links: a unified system for processing faces in the macaque temporal lobe. Science 320 (5881), 1355–1359.
Motter, B.C., 1993. Focal attention produces spatially selective processing in visual cortical areas V1, V2, and V4 in the presence of competing stimuli. J. Neurophysiol. 70, 909–919.
Motter, B.C., 1994. Neural correlates of attentive selection for color or luminance in extrastriate area V4. J. Neurosci. 14, 2178–2189.
Munuera, J., Rigotti, M., Salzman, C.D., 2018. Shared neural coding for social hierarchy and reward value in primate amygdala. Nat. Neurosci. 21 (3), 415–423.
Ni, A.M., Maunsell, J.H.R., 2019. Neuronal effects of spatial and feature attention differ due to normalization. J. Neurosci. 39 (28), 5493–5505.
Piëch, V., et al., 2013. Network model of top-down influences on local gain and contextual interactions in visual cortex. Proc. Natl. Acad. Sci. U. S. A. 110 (43), E4108–E4117.
Poggio, G.F., 1974. Central neural mechanisms in vision. In: Mountcastle, V.B. (Ed.), Medical Physiology, thirteenth ed. C.V. Mosby, Saint Louis, MO, pp. 497–536.
Roy, A., Shepherd, S.V., Platt, M.L., 2014. Reversible inactivation of pSTS suppresses social gaze following in the macaque (Macaca mulatta). Soc. Cogn. Affect. Neurosci. 9 (2), 209–217.
Sadagopan, S., Zarco, W., Freiwald, W.A., 2017. A causal relationship between face-patch activity and face-detection behavior. elife 6, e18558.
Shepherd, S.V., Freiwald, W.A., 2018. Functional networks for social communication in the macaque monkey. Neuron 99 (2), 413–420.e3.
Shepherd, S.V., Platt, M.L., 2008. Spontaneous social orienting and gaze following in ring-tailed lemurs (Lemur catta). Anim. Cogn. 11 (1), 13–20.
Sliwa, J., Freiwald, W.A., 2017. A dedicated network for social interaction processing in the primate brain. Science 356, 745–749.
Treue, S., Martinez Trujillo, J.C., 1999. Feature-based attention influences motion processing gain in macaque visual cortex. Nature 399, 575–579.
Volkow, N.D., Michaelides, M., Baler, R., 2019. The neuroscience of drug reward and addiction. Physiol. Rev. 99 (4), 2115–2140.
Watson, K.K., et al., 2015. Genetic influences on social attention in free-ranging rhesus macaques. Anim. Behavior. 103, 267–275.

Further reading

Hubel, D.H., Wiesel, T.N., 1968. Receptive fields and functional architecture of monkey striate cortex. J. Physiol. 195 (1), 215–243.

Kanwisher, N., Yovel, G., 2006. The fusiform face area: a cortical region specialized for the perception of faces. Philos. Trans. R. Soc. Lond. Ser. B Biol. Sci. 361 (1476), 2109–2128.

McMahan, U.J., 1990. Steve: Remembrances of Steven W. Kuffler. Sinauer, Sunderland, MA.

Pettine, W.W., et al., 2019. Laminar segregation of sensory coding and behavioral readout in macaque V4. PNAS 116 (29), 14749–14754.

Tsao, D.Y., et al., 2003. Faces and objects in macaque cerebral cortex. Nat. Neurosci. 6 (9), 989–995.

CHAPTER 6

Auditory add-ons during vertebrate evolution

Social encounters often start with speech: a question, a call, a familiar greeting. Unlike a baby's cry, which is at once universal but vague in meaning, speech is culture-bound but immediately intelligible. Hearers know what the initiator meant virtually at the speed of sound; they can respond with equal speed, or as quickly as they can formulate a desired response. Speech is a highly effective means of inviting a social approach. If we respond to what someone says, we usually do so with more conversation, employing speech that is apprehensible to whomever began the exchange. Speech is efficient because language can be so precise, and our hearing—at least at close range—creates so little distortion.

I am puzzled, therefore, when linguists ask "why do we have language?" Language encodes and externalizes our thoughts. We can make our feelings *known*, through phonation and the hearer's auditory and related brain mechanisms, designed for that purpose. But for human speech, our social interaction would be stifled or, at best, inescapably primitive. Language operates as social signal and then as mediator, allowing others to enter our mental space and form bonds of every conceivable kind. It also allows others to deliver threats and allows us to answer, sometimes saving us from perpetrating mayhem. It also helps enforce the rules since it's better to talk than commit murder. Though there are thousands of languages, it is no surprise therefore that the capacity for language is evident in every human community. Humans have the combined capacities to generate complex sounds with extraordinarily complex meanings that can be heard and interpreted so accurately.

There is considerable debate concerning when language first developed. Estimates vary wildly from 50,000 years ago to over 2 million. One recent theory suggests that language was coincident with the rise of tool-making, since proto-humans could learn more easily when demonstrations of the art were accompanied by verbal explanations. (In experiments with college students, those who had the benefit of explanation could make simple stone

Origins of Human Socialization
https://doi.org/10.1016/B978-0-323-85861-8.00013-X

tools more readily than those who did not.) In any case, what form these early languages took and how far they varied from mere gesturing is unknown. In the 18th century, *Gulliver's Travels* satirized notions that complex linguistic rules could be abolished, freeing language to be simple and direct. If the means of explaining early tool-making were pretty simple, then by the time of *Gulliver* language had long since evolved to be among the most complex of cultural artifacts. As we know now, language only gets more complex, and is constantly changing to reflect our mores, tastes, commerce, and technology. Purists object to any departures from a supposed linguistic sublime (take, for example, the French Academy) but, as Samuel Johnson observed in his great 18th century *Dictionary of the English Language*, "tongues, like government, have a natural tendency to degeneration." Fortunately, I am not concerned with language as the product of any particular culture, but with language per se—i.e., with physical speech—and how it is processed in the brain.

My concern with speech and how it is interpreted, is a counterpart to my interest in vision and how images are recognized. This is because a response to what other people say—like a response to another person's image—is a social act. Accordingly, for this book, the exact nature of any response is less important to me than the fact that we respond. I am interested in language as a stimulus to action. That our brains are wired to process language suggests that they have *evolved* to do so because language—and the prosocial acts with which we respond to it—are instrumental to the species' survival. We should regard our language capacity similarly to how we regard our capacity to recognize faces: both serve purposes that are essentially protective, stimulating social exchanges in which individuals can work together notwithstanding a challenging environment.

And of course, the circuitry in our brains that allows us to speak and interpret speech is therefore one of the brain mechanisms that I cite as a strand in my three-part theory concerning the origin of human sociality. Along with evolution and babies' development (related to speech) our auditory/interpretive apparatus helps to make sociality efficient and biologically adaptive.

This chapter covers a few of the basics of how speech is produced and heard. While I start with "human" speech, this book recognizes that sociality is broadly distributed among vertebrates (remember the chapter on fish). Because I approach speech generically, rather than as a cultural artifact, I will include some interesting data concerning animals' communication abilities. My point is to situate human speech within a broad evolutionary scheme that has made possible the cohesion of animal—and human—groups by the use of complex auditory signaling capabilities.

In the course of my reading, two surprises blew me away: proofs of active listening (the brain controlling its own auditory pathways); *and* a gradualistic (i.e., evolutionary) genetic approach to human linguistic capacities that winds its way through birdsong among other non-human signaling practices.

Let's start with the production of speech and then move to active listening.

Human speech

Initial thinking about language production started with the human cerebral cortex. Historically, the discovery by the 19th century French anatomist Pierre Paul Broca that injury to part of the left frontal lobe left patients unable to speak, led to the definition of "Broca's area" as the site in the brain for producing speech. But as emphasized by Hugh Duffau (2018), neurosurgeon in Montpellier, Broca's area is not "the speech center," and instead must be viewed as part of at least one network for producing speech. Of special importance for speech near Broca's area is the human laryngeal motor cortex, which makes sense since our larynx is crucial for producing normal speech (Dichter et al., 2018).

So, where do the neural networks, essential for speech, actually go? Within the cortex there are two main streams of axons connecting the frontal cortex, including Broca's area, with the posterior temporal cortex. These two streams, one more ventral and the other more dorsal, have recently been characterized by Ries et al., (2019). One interpretation of their separate functions suggests that the dorsal stream would primarily provide the motor commands for speech, while the ventral would regulate semantic performance. Another interpretation envisioned "the dorsal stream responsible for combining elements into a sequence and the ventral stream for forming semantic dependencies independent of sequential order." Ries' data from studies involving fiber tract disruption indicate "that the dorsal pathways are critical for organizing words in a sequence necessary for sentence generation, and the ventral pathways are critical for processing semantics," that is, for attributing meanings to language. Chang et al. (2015) conceive of the two streams differently, reviewing findings that "The dorsal pathway is linked to sensor-motor mapping of sound to articulation and auditory feedback control in speech production making it important for supporting repetition." By contrast, the ventral stream has been associated with the ability to map auditory input onto conceptual and semantic representations, as well as syntactic processing."

Genes expressed in cortex, and connected to language production

Other cortical axons important for speech descend to the basal ganglia, which are extra-cortical motor control regions of neurons in the basal forebrain. Failures of this network are responsible, for example, for impairment of speech in Parkinson's disease (Manes et al., 2018). Van Rhijn and his colleagues at the University of Nijmegen in the Netherlands (2018) reflected that in children, mutations of a so-called forkhead gene, FoxP2, can cause speech and language deficits and that "loss of Foxp2 in mice is known to affect striatal development and impair motor skills."

Chapter 11 will explain Forkhead-box protein 2, *Foxp2*, a gene transcription factor of great importance during development of the brain. These genes produce proteins important for cell growth, cell proliferation and, ultimately, organogenesis. For example, Kumamoto and Hanashima (2017) found that forkhead box protein family, Foxg1 is crucial for forebrain ("telencephalic") development by commanding downstream cascades of gene expression.

What Nijmegen's Van Rhijn (2018) did was to study the balance between excitatory and inhibitory pathways through the striatum, a balance which I have theorized as crucial for the regulation of social approach (Pfaff and Barbas, 2019) and which becomes unbalanced during the development of autism. Van Rhijn "showed that heterozygous loss of Foxp2 decreases excitatory (AMPA receptor-mediated) and increases inhibitory (GABA receptor-mediated) currents in D1 dopamine receptor positive MSNs of juvenile and adult mice. Furthermore, reduced Foxp2 expression increases GAD67 expression, leading to both increased presynaptic content and release of GABA." In children, you might expect this altered mechanism to lead to disorders of language production.

The FoxP2 gene also played a central role in the work of Schreiweis et al. (2019). Since Schreiweis knew that several types of FOXP2 mutations are associated with speech problems, he genetically engineered a mouse to carry two human-specific amino acid substitutions in the resulting FOXP2 protein and found that such a mouse ($Foxp2^{hum/hum}$) shows profound changes in striatum-dependent behavior and neurophysiology. In line with the Van Rhijn work just cited, these results point to cortical efferents through the basal ganglia, *dependent on FOXP2 genetic signaling*, are important for speech production.

Some of these basal ganglia neurons have axons which loop back toward the cortex through the thalamus, especially the central thalamus. Consider the results of Bhatnagar and Mandybur (2005) which "revealed that the stimulation of the left centromedianum nucleus (in the thalamus) had resulted in a notable facilitatory influence on the processing of language functions; this facilitatory effect was measured in terms of the number of errors produced and the patients' response latency time." Since Bhatnagar (and I) consider the centromedianum and other parts of the central thalamus essentially to constitute a forward extension of the brainstem reticular formation, one might treat these results as demonstrating effects of elevated generalized CNS arousal and alertness (reviewed, Pfaff, 2006, 2019) dependent on the reticular formation.

Brainstem

What happens when speech signaling leaves the forebrain? Some researchers credit the midbrain periqueductal gray (central gray) with having a role in initiating speech. However, there is no way that this tiny area with multiple functions across a range of instinctive behaviors can supply the neuronal information necessary for speech articulation. Rather, we must consider axonal projections from the forebrain to the nerve cell groups in the brainstem which give rise to the cranial nerves, whose axons emerge directly from the brain above the level of the spinal cord.

Clearly important for normal speech production are precise, exquisitely timed signals from the forebrain to four cranial nerves: in order of neuroanatomical emergence from the brain, they are the "fifth cranial nerve" for jaw movements, the seventh for facial movements, the ninth for control of the pharynx (Fig. 1), and the twelfth for control of tongue movements. Lack of accuracy in any one of these brainstem systems will cause speech deficits.

Of course, as a neuroscientist, I naturally emphasize neuronal mechanisms of speech production from cortex through basal ganglia through brainstem and cranial nerves. But why (looking at the whole person) do we have this specific configuration of nerves, muscles, and organs? For the answer, we must also look to evolutionary developments in the anatomy of the mouth and face. According to science writer Bruce Bower, linguist Charles Hockett suggested more than 30 years ago that when humans started farming, they shifted to softer foods which altered their jaw anatomy, allowing them to make sounds like "f" and "v." So, should the evolution of speech

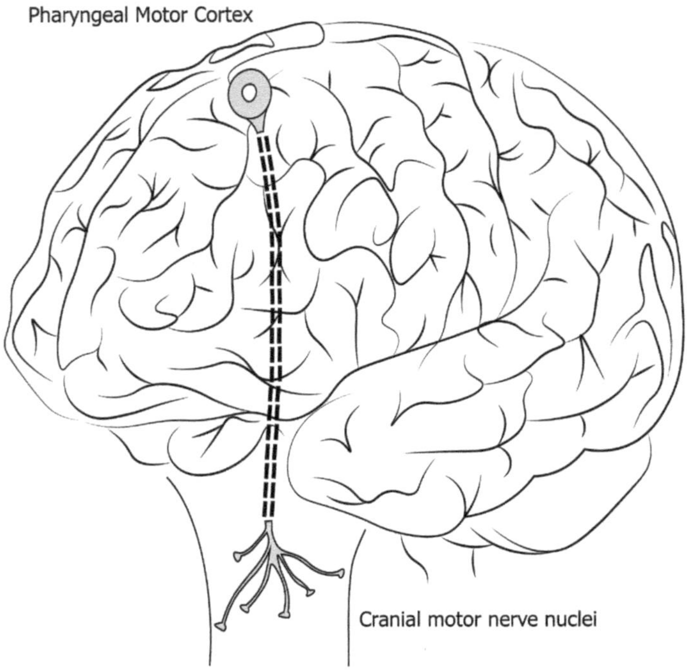

Fig. 1 Axonal projections from "pharyngeal motor cortex" which help to regulate sound production provide a unifying concept for understanding vocal learning in various species.

production be deduced from teeth? Blasi et al. (2019) tested this idea and inferred that "labiodental sounds (such as 'f' and 'v') were… due to changes in diet attributable to food-processing technologies modified the human bite from an edge-to-edge configuration to one that preserves adolescent overbite and overjet into adulthood." The important lesson from this and other work is that some linguistic developments stem from the anatomical influence of economic and hence cultural change.

Thus, while human-specific language skills have long focused the attention of neuroscientists, there is still a vast amount to learn regarding how speech production developed. A relevant perspective for this book is that speech as a physical capability presumes, and has evolved because of the proximity of others. Human speech can impose organization, even as it has evolved to reflect social organization. Thus, we must be able to interpret others' speech, not just emit speech with our lungs after making words with our brains, muscles, and facial organs. We have to hear what other people are saying—initially, and in response to our claims.

Hearing

On the other side of that initial, potentially friendly social exchange, how do we hear? Compared to speech production mechanisms, the mechanisms for hearing are much better understood. In this chapter, I cover the basics, since my main concern is with the interpretation of what we hear and with our response.

To begin, the simplest way to begin discussing how a person responds to a social greeting (or anti-social threat) is to describe the physics of sound up to the point that it reaches the brain, i.e., the reception of sound in the inner ear (Goldstein, in Mountcastle, 1974, p. 391, ff.), and the neuroanatomy and neurophysiology of auditory pathways in the brain.

Briefly, the sound waves caused by a person's voice—"successive compressions and rarefactions of air particles" (p. 386)—agitate fluid in the ear, causing vibrations of membranes in the inner ear; these vibrations cause movements of hairs on the receptor cells of audition.The forces on these cells lead to generation of electrical signals in neuronal fibers which will travel in the eighth cranial nerve into the hindbrain.

Before things get complicated, consider the neural/auditory activity in the brain itself. In the hindbrain, auditory fibers arrive at a 3-part cell group called the cochlear nuclei, where the auditory information is *tripled* by 3-way orderly distributions of incoming nerve fibers across three cochlear nerve cell populations (Mountcastle, 1974, p. 421). As we move up the auditory system, there will be many fibers crossing between left and right sides of the brain.This is important because the exact timing of sounds on the left and the right gives the listener cues as to the direction of the voice.Auditory information is forwarded up the brainstem in a "highly ordered functional topography" (p. 424) until the signals reach two very different nerve cell regions in the upper brainstem. First, signals will reach layers of cells resident in a "bump" on top of the brainstem, the inferior colliculus (Latin word for "little hill").This information is not forwarded to the cerebral cortex. Second, voice signals will reach a part of the thalamus called the medial geniculate body. Some of these medial geniculate neurons are essential for forwarding the processed auditory signal to the cerebral cortex.

Sound representations in the primary auditory cortex, in the temporal lobe of the human brain, are beautifully organized, again in triplicate, according to the frequency of each component of the sound.The central core of sound processing in this primary region "Auditory I" shows neurons most responsive to low frequencies (about 500 Hz) in the anterior temporal lobe,and neurons most responsive to high frequencies (20,000–25,000 Hz) are found in posterior temporal cortex.In turn, primary auditory cortex is surrounded by other cortical regions whose roles in receiving speech are less well defined.

For social responses to social signals, neurons which respond specifically to voices would be particularly exciting to understand. As Chris Petkov stated, functional magnetic resonance imaging (fMRI) studies localized voice-preferring regions in the primate temporal lobe, "but the hemodynamic response cannot directly assess neurophysiological properties." He and his team in Tubingen investigated the responses of neurons in an fMRI-identified voice cluster in awake monkeys, and provided the first systematic evidence for voice cells.Voice cells were identified (Perrodin et al., 2011) by way of analogy to face cells in the visual system, and were defined as "neurons responding at least 2-fold more strongly to conspecific voices than to nonvoice sounds or heterospecific voices."

So, this represents the simplest, most straightforward description of how a sound, a greeting or a threat, would be received in the brain.But, in addition, there have been startling developments:the first takes place at the very beginning of the story just presented (in the structure of the ear); and the second takes place at the very end (in the cerebral cortex). I will describe these developments below, and would not be surprised if additional complexities were to emerge at any point in the process.

Active listening

"Active listening" mechanisms range from amplification by hairs in our ears to top-down processes in cerebral cortex analogous to top-down processes for active visual attention in the foregoing chapter. This science goes far beyond passive reception of sound by the auditory pathways. So, first, I talk about hair bundles in the inner ear, and then then I discuss the role of the auditory cortex.

At the very beginning of the auditory process, we benefit from a process that "amplifies acoustic inputs by more than a hundred-fold, [and] sharpens frequency discrimination to facilitate the comprehension of speech" (Hudspeth, 2014). How does this occur? Hair bundles in the inner ear oscillate; these movements are entrained by incoming auditory stimuli; the stimuli are then amplified. Characteristic of certain non-linear dynamic systems, a critical value of a "control parameter" in an exponential equation uses dynamic instability which leads to an abrupt and large change at what is called a "Hopf bifurcation."The control parameter is the name for one term in a non-linear equation affecting hearing. This abrupt change is occasioned by huge amplification of the sound we are listening to. Choe et al. (1998) state: "As an eigenvalue crosses the imaginary axis, an oscillatory decay mode becomes a limit-cycle oscillation (characteristic of some non-linear oscillatory systems. The parameter set at the crossing constitutes a Hopf bifurcation... (which) assures the existence of a regime in which the system can act as a tuned amplifier. Although a hair bundle needs not display sustained spontaneous oscillation, the auditory system might be expected to exploit the local instability."

This is not passive, linear listening. It is active, and near this critical "bifurcating" value of the control parameter in the relevant equation, auditory sensitivity grows and tuning to a specific frequency sharpens. In fact, for smaller stimuli the amplification becomes larger. This amplification shows how the signals coming out of the ear can be actively controlled.

At the top of the auditory circuitry, we run into another example of a proactive system. Again, analogous to vision, *top*-down means control of the auditory cortex from other higher brain regions which themselves deal with high cognitive processes. The notion that expectation, attention and learning affect the processing of auditory signals in the cortex is well accepted. Some of the inputs to auditory cortex which do this come from the frontal cortex. As Yang Dan's team at Berkeley said (Zhang et al., 2016), "Long-range projections from the frontal cortex are known to modulate sensory processing in multiple modalities." Zhang used virus-assisted circuit mapping to identify the brain networks for top-down modulation of visual, as well as somatosensory and auditory processing. The visual cortex is reciprocally connected to the anterior cingulate area, whereas the somatosensory and auditory cortices are connected to the primary and secondary motor cortices. Especially when such top-down connections deal with the left auditory cortex, downstream from Broca's area, they offer the possibility that comparison to speech signaling might foster speech understanding.

Neurons in cortical layers 2/3 and 5 in the primary and secondary motor cortices project back to auditory cortex. This kind of positioning, even in mouse brain, provides the neuroanatomical basis for top-down control of auditory cortical functioning.

Such top-down control, signaling from more advanced sensory processing neurons "backward" to influence the less advanced, offers a multitude of opportunities for sophisticated auditory signal processing. Kuchibhotla and Bathellier (2018) need top-down cortical control in order to use "behavioral context and task-related information" to influence cortical encoding of sounds. Irvine (2018) cites "complex distributed networks" to explain auditory attentional processes, decisions and susceptibility to reward. With respect to reward, on the one hand, auditory signaling in the cortex must have an element of changeability, learning, so-called "plasticity." But on the other hand, that requirement is opposite the need for auditory perceptual stability (Irvine, 2018).

Neurobiologists face another problem at the top of our auditory circuitry in the cortex: how can we hear and unscramble sounds under challenging conditions, how is auditory perception influenced by learning, and, how can we understand speech when there is so much surrounding noise? (Recently, there was a dispute in the Food section of the New York Times, where restaurant-goers took sides over whether excessive noise levels should weigh against the quality of an establishment's food). Park et al. (2015) showed how other, non-auditory parts of the human cerebral cortex can influence auditory signaling. In their results, frontal and motor cortices "significantly modulated the phase of speech-coupled low-frequency oscillations in auditory cortex." These authors suggest that low-frequency brain oscillations play a role in implementing predictive top-down control during continuous speech perception.

Márton et al. (2019) use the phrase "predictive coding" to describe top-down and bottom-up interactions in auditory and other sensory processing. These authors think of a "cortical hierarchy" used in auditory processing. Rather than simply dealing with the reception of elementary auditory stimuli, Márton is concerned with complex auditory signal processing which includes nonlinear messaging and bidirectional information transfer.

One aspect of this information processing, Bayesian (conditional probability) modeling, is invoked by Aller and Noppeney (2019) to explain the brain's formation of an (auditory) percept. These authors argue that the brain accomplishes "Bayesian causal inference": they say the brain's estimate of the probability of an element of information depends on the nature of a previous event. Instead of trying to rely simply on an unsophisticated registration of an auditory or visual stimulus set at the human cerebral cortex, Aller speaks of the "dynamic evolution of perceptual estimates." As I read these authors, their information processing could apply both to the appreciation of sound and to the comprehension of the sound's meaning.

Finally, to obtain evidence for top-down processing, Italian neurophysiologists Marinato and Baldauf (2019) used a complex auditory object identification task which was either aided or not by a visual cue. "The participants' responses were significantly faster and more accurate in the valid cue condition compared to the invalid cue condition, indicating a robust cue-validity effect of high-level, object-based auditory attention." That is, there was top-down processing.

Even music perception (Koelsch et al., 2019) is thought of as "an active act of listening." These authors state that "when listening to music we constantly generate plausible hypotheses about what could happen next, while actively attending to [how] music resolves the ensuing uncertainty." Top-down auditory processing allows for "attentional selection of salient auditory features. This formulation suggests that 'listening' is a more active process than traditionally conceived in models of perception."

Sounds' meaning. Max Planck Institute professor Angela Friederici and her team (2009) have subsumed top-down, active system thinking into their investigations of speech comprehension. While speech strings which she calls "semantic-free grammars" activated only prefrontal cortex in her functional magnetic resonance imaging (fMRI) recordings, complex hierarchically structured sentences maximally activated regions in the temporal lobe of the cortex. As Friederici (2012) writes, both ascending speech signals reaching the auditory cortex and superior temporal cortex and "top-down, controlled and predictive processes from the prefrontal cortex back to the temporal cortex" constitute a cortical language circuit. Friederici and Singer (2015) think that different structural networks in the cortex underlie understanding of several aspects of language. Most surprising, there are single neurons which react in similar ways to a person's name as to the person herself.

Building on this area of linguistic work, Getz and Toscano (2019) used the electrophysiological "event-related-potential technique" to examine the effect of semantic (meaningful) context on the auditory cortex's response to words. For example, experimental subjects "heard auditory targets (e.g., 'potatoes') following associated visual primers (e.g., 'MASHED')," compared to other, irrelevant visual primers. As would be predicted by a cross-sensory interactive top-down model, "semantic context modulated responses to upcoming targets." Understanding spoken words depends on top-down processing. Indeed, the importance of how we process language to arrive at its meaning cannot be over-estimated. If we hear something but cannot understand its implications for us, then we will just stand there unresponsively. There would be no point to our capacity to hear/amplify/refine language if we could not interpret and respond to it. Our brains are, in fact, so sophisticated that we do not just assign dictionary meanings to words that we hear but actually interpret them in context. We have learned to make split-second decisions about words based on how they are said—e.g., in English the same words said in a raised voice implies a query ("You're well?" vs. You're well!"). Our brains apprehend the social circumstances of other people's language production, so that our understanding of what they say is part of a *gestalt* process that extends beyond the literal translation of sounds into words. Our brains are wired to produce meaning.

Accordingly, Lau et al. (2019) agree that "speech processing is highly modulated by context." These authors used an electrophysiological technique called Frequency Following Responses (FFRs) which reflect activity of neuronal ensembles in the auditory system to see if they would be modulated by abstract linguistic characteristics. To keep questions of linguistic familiarity out of the study, they used Mandarin, in particular a syllable which requires a rising tone. This Mandarin syllable was presented in three different tonal contexts. Speech encoding as reflected by FFR was context-sensitive, indicating interactions between bottom-up and top-down mechanisms of speech perception. Lau et al. think that the auditory system "predictively fine-tunes incoming speech encoding using linguistic and statistical information from prior context.

Overall, Sophie Scott, from University College London (2019) characterizes speech as the auditory form of language and splits up the brain's tasks into Voice Processing and Linguistic Processing. The temporal cortex on the right side of our brains deals more with Voice Processing (phoneme recognition, pitch, intonation, talker identification). On the left side, Linguistic Processing (e.g., syllabic structure). And she highlights the need for neural machinery outside the temporal lobe—e.g., premotor cortex—to manage "the swift changes of talker turns that characterize spoken conversations" (p. 61).

Overview. Thus, decades of exploration and analyses of auditory pathways have led to a satisfying clarity of how sound signals eventually reach the cortex. But novel thinking about the dynamics of the system has broken new ground at the front end of the system, in the ear, and at the top of the system, in the cortex. Listening to social sounds as to other sounds is an active process.

Evolution, genetics, and continuous (not dichotomous) developments

Recently gathered genetic data, in addition to good evolutionary thinking, allow us to recognize that human speech is not entirely discontinuous with the development of non-human auditory communication. Rather, across the animal kingdom there have been continuous developments of such communication. It is not true that biology jumped from "no communication" to human speech. In discussing some of this now, my point is to demonstrate that evolution favors auditory communication because it helps diverse groups to cohere according to social rules and thus to survive.

As an undergraduate, I heard psychologists and philosophers talk about humans as the only species capable of language. So, when the great linguist, MIT professor Noam Chomsky wrote *Why Only Us* (2016, with Berwick), I assumed it would celebrate human uniqueness as the only vocal-learning species. After all, I was initially skeptical that birdsong would have much to teach us concerning human speech. But Chomsky adopted an open-minded perspective regarding the search for universal principles of vocal learning, and a non-dichotomous view of developments leading to human language. I came over to his side. Of course, this is not to say that birds—or any animal species—have language capabilities like ours. They cannot produce Johnson's *Dictionary*, much less the *Oxford English Dictionary* (defining 171, 464 current words and 47,156 obsolete ones). They do not have the neuroanatomical apparatus to articulate those words or interpret their meanings. But they do have communication capabilities along a spectrum of capabilities that lead us to infer that they have their own language systems; and that these are immensely useful within their own groups. They also show a surprising similarity with regard to genes that are necessary for vocal learning.

According to Chomsky, people "since the time of Aristotle have been pondering whether birdsong serves as a good model for language" (p. 140). Both modes of communication depend on precise, rapidly operating finite state transition networks. And indeed, at this level—which Aristotle could not even have imagined—there is some plausibility of comparison. We are not talking about words here, or complicated grammar, but the tonalities underlying words which characterize signaling in a multiplicity of forms. In terms of signaling, humans and birds share a common denominator (which is low but not vanishingly low).

With respect to acknowledging the neuroscience and genetics of vocal learning, Chomsky has come a long way since his 1957 theory of a universal "generative grammar," in which he speculated about how the "basic properties" of grammar are generated. In *Why Only Us?* that kind of structural thinking gives way to the physical realities of systems sometimes linked to biological "input/output" features of vocal learning. Of course, Chomsky always had at least a nodding acquaintance with neuroscience insofar as he argued that the human brain is "hardwired" for innate language comprehension. But his newer approach gets closer to the specifics of how brains (human and animal) work in tandem with anatomical structures to make possible what physicists would call signaling. In this regard, Chomsky unifies linguistic capabilities (so defined) across a range of species, facilitating the study of their shared evolution.

So while I began by expecting Chomsky to argue the absolute exclusivity of human language, I was interested to see him focus on neurobiologist Erich Jarvis' view (Pfenning et al., 2014) that expression of perhaps 100–200 genes are crucial for "the ability to learn distinctive, ordered sounds" and that these genes work through expression in neuronal pathways that can be seen as parallel in songbird and human brains. With an emphasis on inputs and outputs essential for vocal learning, Chomsky is concerned with the developmentally expressed gene FOXP2 for developing the sensorimotor interface necessary for verbal learning. Mutations in FOXP2 are associated with a heritable language defect called verbal dyspraxia (which includes difficulty in making speech sounds and in sequencing sounds) as well as other motor deficiencies including "deficits in learning of rapid motor skills"; the gene's mutations are associated with vocal performance, not with "the core faculty of human language" (p. 78). The FoxP2 mutation affects speech production, not its understanding. So, we return to the forkhead gene family. I will further discuss Fox/forkhead gene families below and in Chapter 11.

The studies most heavily cited by Pfenning et al. (2014) suggest that song-learning birds and humans share independently evolved similarities in brain pathways for vocal learning that are essential for song and speech and are not found in most other species. Comparisons of brain transcriptomes of song-learning birds and humans relative to vocal non-learners identified convergent gene expression specializations in specific song and speech brain regions of avian vocal learners and humans. The strongest shared profiles relate bird motor and striatal song-learning nuclei, respectively, with human laryngeal motor cortex and parts of the striatum that control speech production and learning. Most of the associated genes function in motor control and brain connectivity. Thus, convergent behavior and neural connectivity for a complex trait are associated with convergent specialized expression of multiple genes.

Accordingly, given the genetic convergence, we can ask: what is the practical influence on speech of FoxP2, a member of the so-called "forkhead" family of gene transcription factors? These forkhead factors control the expression of important regulatory genes and are crucial for normal development of the brain. Pondering this question, I turned to my Rockefeller colleagues, molecular biologists Robert Roeder and James Darnell. I learned that Clark et al. (1993) determined the three-dimensional structure of a DNA sequence that accommodates binding by a forkhead transcription factor. Darnell's lab (Lai et al., 1993) described "the already known sequence of a Drosophila nuclear protein termed fork head. The fork head gene was so named because embryos mutant in the gene were defective in the formation of terminal (end) structures that normally give rise to the anterior and posterior gut, resulting in two spiked head structures." Because of the defects in the formations of organs caused by forkhead protein mutations, the gene family, in Darnell's words, "caught the interest of developmental biologists."

Friedman and Kaestner (2006) agree: "The *forkhead* box (*Fox*) gene family is named for the *Drosophila* gene *fork head*, mutations in which cause defects in head fold involution, resulting in a characteristic spiked head appearance. The FORK HEAD protein contains a 110-amino acid motif that is conserved from yeast to man and which functions as a DNA-binding domain. Over 100 *Fox* genes have been identified and classified into subfamilies, and many have been shown to have important biological functions by genetic analyses." Of course, most interesting to neuroscientists will be the contributions of Fox genes to organogenesis, because we are looking for new developments in the human cerebral cortex to provide mechanisms for intelligible human language.

As a neurobiologist, these developments tell me that FoxP2 does not contribute to language production in a specific and targeted manner. Instead, such mutations disrupt neural circuitry, perhaps in a widespread manner, thus interfering with a complex behavioral function that for its performance needs everything to work perfectly. *A complex behavioral function is easier to mess up than a simple behavioral function.* Likewise, axon guidance gene products such as "SLIT1" and "ROBO1" are not dedicated to language. Instead, they provide the substrate for a wide variety of behavioral functions; vocal learning places a high demand for coordinated neuronal performance, a level of performance which is easily interrupted. Again, how easy is it to interrupt complex, precisely timed productions, compared to simple processes? An NFL football play can be interrupted because it involves 11 people performing under duress with split-second timing. The football play is more complex than the 100-yard dash. Computers are likewise susceptible because their operations depend on the precise operation of hundreds of chips which, in turn, are bundles of transistors. A complex machine like a grandfather clock is easier to damage than a simple lever.

In our present case of language production, vocal learning is susceptible to genetic mutations of various sorts and neuronal deficits however small, exactly because vocal learning is a highly informational process with requirements for temporal and muscular precision of the highest order.

All of these considerations were recognized by Pfenning and Jarvis (p. 1333), who wrote that "The finding that convergent neural circuitries are accompanied by convergent molecular changes of multiple genes… indicates that brain circuits for complex traits may have (only) limited ways in which they could have evolved." Jarvis recognized, by analogy, that complex traits like verbal learning are more like an NFL football play than a 100-m dash in that they were more difficult to produce and easier to disrupt. While the most categorical difference between vocal learning species and non-learners (Jarvis, 2019) is the presence of a neuronal pathway (in the former) from laryngeal motor cortex to vocal motor neurons, Jarvis lists several quantitative differences as well.

Greg Gedman, with Erich Jarvis, paints a picture of molecular convergences associated with behavioral convergences, comparing vocal learning performance (the production of sounds via the imitation of a model) in songbirds and humans. Parallel features between songbirds and humans include imitation, a critical period in life during which learning occurs, and the inevitable presence of abnormalities. Networks involved include projections from cerebral cortex to basal ganglia with loops back into the thalamus. The parallels also include laryngeal motor cortex projecting to hind brain and subcortical paths projecting to hindbrain. Parallels in gene expression show up in laryngeal motor cortex. In both birdsong and human speech, prominent features include fine motor control with exquisite timing.

A new report supports Erich Jarvis' view of continuous, non-dichotomous evolutionary routes toward development of the brain mechanisms for human speech. As summarized by Steffen Hage in Science (2019), "mice, monkeys and humans have a phylogenetically conserved primary vocal motor network that is capable of coordinating all motoneuron pools involved in vocalization" (p. 926). These move forward in evolutionary time from a relatively simple control network to human networks which are highly elaborate. For humans, the control network is Broca's area in the left cerebral cortex, coupled with primary motor cortex, premotor and prefrontal cortex, basal ganglia, the midbrain periaqueductal gray matter, and brainstem cranial motor nerve nuclei. Hage's piece was stimulated by the discovery from Michael Long's New York University lab (Okobi et al., 2019) that a species of mouse displayed social singing behavior which requires control by neurons in an orofacial motor control region.

Further evidence for Jarvis's non-dichotomous approach to the evolution of spoken language derives from an analysis of the neural components involved and from a consideration of exactly what kind of message is delivered. Birds have birdsong for obvious biological purposes such as territorial control in the service of mating. First, in terms of the neural systems involved, we are not talking about a sudden jump from a species that does not use vocalizations to a species whose exemplars include Churchill and FDR. Instead, consider what Chis Petkov and Erich Jarvis (2012) call "motor theories." These posit that "the brain pathways that control the learning and production of song and speech were derived from adjacent motor brain pathways." In both lower animals and in humans there are forebrain systems. What may differ between species are *continuous* variables such as the numbers of neurons on the job, their sensitivities, their dynamic ranges, the distributions of their synaptic hookups, and the details of laryngeal controls. Second, consider the hierarchy of the communicative behaviors to be controlled. In speech we have a *temporal sequence* of sounds. At a more abstract level, we have *syntax* (rules for sequencing parts of speech to arrive at sentence structures). Finally, there is *semantics* (the interpretation of speech elements in order to understanding meanings).

In order to help understand the evolution of linguistic ability, Greg Petkov cleverly avoided more arcane subjects, which we skip by here, to focus on sequence processing (Milne et al., 2016). At the Institute of Neuroscience, and Centre for Behavior and Evolution of Newcastle University, Petkov wanted to analyze "neurobiological processes at specific human development stages," to see how they compare with those in phylogenetically closely related species. They measured EEG event-related potentials (ERPs) in two macaque monkeys using the same experimental paradigm as that developed to evaluate human infant and adult brain potentials associated with the processing of non-adjacent (i.e., more difficult) ordering relationships in sequences of syllable triplets (Milne et al., 2016). "Frequent standard triplet sequences were interspersed with infrequent voice pitch or non-adjacent rule deviants. Monkey ERPs show early pitch and rule deviant mismatch responses that are strikingly similar to those previously reported in human infants." The results speak to evolutionarily conserved neurophysiological effects, some of which are remarkably like those seen at an early human developmental stage. By analyzing sequence, Petkov and his team took a step toward explaining evolutionary conservation of mechanisms related to syntax.

Having taken this step, the Newcastle team Wilson et al. (2013) could move into the construction and analysis of artificial grammars designed to emulate aspects of the structure (sequence) of language. They centered on an under-studied artificial grammar, one with a forward-branching structure designed to model certain aspects of the nondeterministic nature of word transitions in natural language. Their results "showed that common marmosets (New World monkeys) differentiated between well formed, correct testing sequences and those violating the AG structure based primarily on simple learning strategies. By comparison, Rhesus macaques (Old World monkeys) showed evidence for deeper levels of artificial grammar learning." This provides evidence for some points on a non-dichotomous development of auditory mechanisms serving the understanding of language.

Taking the artificial grammar strategy to the next level, Chris Petkov and his team faced the fact that natural languages have multiple levels of complexity. So, Wilson et al. (2015) tested humans and Rhesus macaques using a mixed-complexity auditory artificial grammar "containing both adjacent (local) and non-adjacent (longer-distance) relationships." Following exposure to exemplary sequences generated by the AG, humans and macaques were individually tested with sequences that were either consistent with the AG or violated specific adjacent or non-adjacent relationships. Petkov's complex set of results showed that the monkeys did quite well with a degree of matching to human performance for the easy (adjacent) relationships; for the harder (non-adjacent, longer distance) relationships, monkey performance fell down, compared to humans.

Overview

While I am passionately interested in human social communications, I recognize that vocal communications are widespread among vertebrate animals. As the evolutionary biologist Roger Payne discovered, whales sing. Diane Reiss reported that dolphins show remarkable vocal abilities from infancy through adulthood. Even some mice sing (Okobi et al., 2019). But we enter a new stage when we escalate the discussion to vocal learning. Obviously, humans learn to use language, and as generations of behavioral biologists have shown, so do songbirds. What's new comes from the work of Erich Jarvis, who emphasizes that the mechanisms underlying vocal learning—genetic and neurophysiological—did not emerge all at once. Rather, continuous, non-dichotomous developments led to speech as we know it.

In the future, people without voices may manage vocal social approaches in an entirely new way. Professor Edward Chang, a neurosurgeon at UCSF (Anumanchipalli et al., 2019), and his team have invented a method for taking signals from the relevant parts of the human cerebral cortex—the ventral sensorimotor cortex, the superior temporal gyrus and the inferior frontal cortex—and using them to synthesize audible speech. They envision "speech neuroprosthetic technology to restore spoken communication" to mute patients. As always, natural evolution continues into machinery, which can simulate, advance, and speed up its processes. In effect, we are starting to reverse engineer evolution to preserve the capacity for speech when it is lost due to natural or traumatic occurrences.

I ended this chapter with a reference to evolution because speech (including its animal counterparts) is a quintessentially adaptive instrumentality allowing species to survive. Without it, groups become mere scattered individuals with much reduced capabilities for cooperation and defense. Moreover, the evolution of brain mechanisms for human speech demonstrates how the three strands of my argument for human sociality merge: not only does it concern *human evolution* and *babies' development* of speech capabilities, but also the resultant, specific *brain mechanisms* for vocal learning.

Clearly, language production and interpretation support human sociality, just as vision does. Without these two highly refined senses, humans would be a vastly inferior species. Our capacity to learn and to pass on knowledge would be limited. At a deeper level, our sense of time would be radically altered, since we can recognize images and language in an instant without having to laboriously piece together random clues with our remaining senses. It is difficult for most of us to imagine a world that we could not see *and* not hear. These two senses work in tandem to provide perspective and add dimension. We know what and where something is, and we can judge its approach or withdrawal. We have learned to enhance both senses with microscopes, telescopes, and microphones. But we are fully human only because these senses are highly evolved and calibrated to the world that we meet at eye-level and within earshot.

Helen Keller (1880–1968) learned to navigate the world and, indeed, to become an extraordinary actor on the world stage, without either sight or hearing. We should all be in awe of her accomplishments. Even the briefest study of her life demonstrates the immense obstacles that she had to overcome. Senses that we take for granted were entirely beyond her.

To appreciate her accomplishment, therefore, is to appreciate how sight and hearing (acting together) connect us to each other and the environment. We now have schools for the blind and for the deaf, but these teach their students how to rely on senses that they do possess. Such compensation, though immensely helpful, is not ideal—an immense amount of time is spent just learning how to compensate. The fact about our sensorium is that it allows us to be quick right from the start, as babies, and all through life to respond to cues that do not have to be reinterpreted by other senses. This quickness favors our survival.

The next chapter will discuss the role of "attachment theory" in modern neuroscience, focusing on how early attachments to others influence human development.

New questions

1. How do the discoveries about speech mechanisms and auditory pathways in the brain inform future development of voice recognition software? To date, such software is understood as fast, but not 100% accurate.

References

Aller, M., Noppeney, U., 2019. To integrate or not to integrate: temporal dynamics of hierarchical Bayesian causal inference. PLoS Biol. 17 (4), e3000210.

Anumanchipalli, G.K., Chartier, J., Chang, E.F., 2019. Speech synthesis from neuronal decoding of spoken sentences. Nature 568, 493–498.

Berwick, R.C., Chomsky, N., 2016. Why Only Us: Language and Evolution. The MIT Press, Cambridge.

Bhatnagar, S., Mandybur, G.T., 2005. Effects of intralaminar thalamic stimulation on language functions. Brain Lang. 92, 1–11.

Blasi, D.E., et al., 2019. Human sound systems are shaped by post-Neolithic changes in bite configuration. Science 363 (6432). pii: eaav3218.

Choe, Y., Magnasco, M.O., Hudspeth, A.J., 1998. A model for amplification of hair-bundle motion by cyclical binding of Ca^{2+} to mechanoelectrical-transduction channels. Proc. Natl. Acad. Sci. U. S. A. 95 (26), 15321–15326.

Clark, K.L., et al., 1993. Co-crystal structure of the HNF-3/fork head DNA-recognition motif resembles histone HS. Nature 364, 412–419.

Dichter, B.K., et al., 2018. The control of vocal pitch in human laryngeal motor cortex. Cell 174 (1), 21–31.e9.

Duffau, H., 2018. The error of Broca: from the traditional localizationist concept to a connectional anatomy of the human brain. J. Chem. Neuroanat. 89, 73–81.

Friederici, A.D., 2012. The cortical language circuit: from auditory perception to sentence comprehension. Trends Cogn. Sci. 16 (5), 262–268.

Friederici, A.D., Singer, W., 2015. Grounding language processing on basic neurophysiological principles. Trends Cogn. Sci. 19 (6), 329–338.

Friederici, A.D., Makuuchi, M., Bahlmann, J., 2009. The role of the posterior superior temporal cortex in sentence comprehension. Neuroreport 20 (6), 563–568.

Friedman, J., Kaestner, K., 2006. The Foxa family of transcription factors in development and metabolism. Cell. Mol. Life Sci. 63, 2317–2328.
Getz, L.M., Toscano, J.C., 2019. Electrophysiological evidence for top-down lexical influences on early speech perception. Psychol. Sci. 24. 956797619841813.
Hage, S.R., 2019. Precise vocal timing needs cortical control. Science 363, 926–928.
Hudspeth, A.J., 2014. Integrating the active process of hair cells with cochlear function. Nat. Rev. Neurosci. 15 (9), 600–614.
Irvine, D.R.F., 2018. Auditory perceptual learning and changes in the conceptualization of auditory cortex. Hear. Res. 366, 3–16.
Jarvis, E.D., 2019. Evolution of vocal learning and spoken language. Science 366, 50–55.
Koelsch, S., et al., 2019. Predictive processes and the peculiar case of music. Trends Cogn. Sci. 23 (1), 63–77.
Kuchibhotla, K., Bathellier, B., 2018. Neural encoding of sensory and behavioral complexity in the auditory cortex. Curr. Opin. Neurobiol. 52, 65–71.
Kumamoto, T., Hanashima, C., 2017. Evolutionary conservation and conversion of Foxg1 function in brain development. Dev. Growth Differ. 59 (4), 258–269.
Lai, E., et al., 1993. Hepatocyte nuclear factor 3/fork head or "winged helix" proteins: a family of transcription factors of diverse biologic function. Proc. Natl. Acad. Sci. U. S. A. 90, 10421–10423.
Lau, J.C.Y., Wong, P.C.M., Chandrasekaran, B., 2019. Interactive effects of linguistic abstraction and stimulus statistics in the online modulation of neural speech encoding. J. Percept Psychophys. 81 (4), 1020–1033.
Manes, J.L., et al., 2018. Altered resting-state functional connectivity of the putamen and internal globus pallidus is related to speech impairment in Parkinson's disease. Brain Behav. 8 (9), e01073.
Marinato, G., Baldauf, D., 2019. Object-based attention in complex, naturalistic auditory streams. Sci. Rep. 9 (1), 2854.
Márton, C.D., et al., 2019. Signature patterns for top-down and bottom-up information processing via cross-frequency coupling in macaque auditory cortex. eNeuro 6 (2). pii: ENEURO.0467-18.2019.
Milne, A.E., et al., 2016. Evolutionary origins of non-adjacent sequence processing in primate brain potentials. Sci. Rep. 6, 36259.
Mountcastle, V.B., 1974. Medical Physiology, thirteenth ed. Mosby, Saint Louis, MO.
Okobi, D.E., et al., 2019. Motor cortical control of vocal interaction in neotropical singing mice. Science 363 (6430), 983–987.
Park, H., et al., 2015. Frontal top-down signals increase coupling of auditory low-frequency oscillations to continuous speech in human listeners. Curr. Biol. 25 (12), 1649–1653.
Perrodin, C., et al., 2011. Voice cells in the primate temporal lobe. Curr. Biol. 21 (16), 1408–1415.
Petkov, C.I., Jarvis, E.D., 2012. Birds, primates, and spoken language origins: behavioral phenotypes and neurobiological substrates. Front. Evol. Neurosci. 16 (4), 12.
Pfaff, D.W., 2006. Brain Arousal and Information Theory: Neural and Genetic Mechanisms. Harvard University Press, Cambridge, MA.
Pfaff, D.W., 2019. How Brain Arousal Mechanisms Work: Paths Toward Consciousness. Cambridge University Press, Cambridge.
Pfaff, D.W., Barbas, H., 2019. Mechanisms for the approach/avoidance decision applied to autism. Trends Neurosci. 42, 448–461.
Pfenning, A.R., et al., 2014. Convergent transcriptional specializations in the brains of humans and song-learning birds. Science 346 (6215), 1256846.
Ries, S.K., et al., 2019. Roles of ventral versus dorsal pathways in language production: an awake language mapping study. Brain Lang. 191, 17–27.

Schreiweis, C., et al., 2019. Mice carrying a humanized Foxp2 knock-in allele show region-specific shifts of striatal Foxp2 expression levels. Cortex. pii: S0010-9452(19)30027-9.
Scott, S.K., 2019. From speech and talkers to the social world; the neural processing of human spoken language. Science 366, 58–62.
van Rhijn, J.R., et al., 2018. Foxp2 loss of function increases striatal direct pathway inhibition via increased GABA release. Brain Struct. Funct. 223 (9), 4211–4226.
Wilson, B., et al., 2013. Auditory artificial grammar learning in macaque and marmoset monkeys. J. Neurosci. 33 (48), 18825–18835.
Wilson, B., Smith, K., Petkov, C.I., 2015. Mixed-complexity artificial grammar learning in humans and macaque monkeys: evaluating learning strategies. Eur. J. Neurosci. 41 (5), 568–578.
Zhang, S., et al., 2016. Organization of long-range inputs and outputs of frontal cortex for top-down control. Nat. Neurosci. 19 (12), 1733–1742.

Further reading

Banerjee, A., et al., 2019. Singing mice. Curr. Biol. 29 (6), R190–R191.
Friederici, A.D., et al., 2006. The brain differentiates human and non-human grammars: functional localization and structural connectivity. Proc. Natl. Acad. Sci. U. S. A. 103 (7), 2458–2463.
Hudspeth, A.J., 2005. How the ear's works work: mechanoelectrical transduction and amplification by hair. C. R. Biol. 328, 155–162.
Hudspeth, A.J., 2008. Making an effort to listen: mechanical amplification in the ear. Neuron 59 (4), 530–545.

CHAPTER 7

Psychological approaches

A vast array of studies regarding human social behaviors are not based in neuroscience. We could go back to the Greeks. Beginning in the 18th century, I can think of Adam Smith, whose *Inquiry into the Nature and Causes of the Wealth of Nations* (1776) explains how people behave in a market economy. Few environments are more intensely social than a market, a perpetual fugue of opportunism, cheating, give, and take. Smith famously claimed that if everyone just went about their business and followed their own best interests, an "invisible hand" would ensure that the market functioned optimally. Of course, we now know that this is an unlikely outcome, but Smith is still cited for his notion that self-interest is a basic behavioral motivation. Also in the 18th century, David Hume's *Theory of Moral Sentiments* (1759) considered the origins not just of our "passions" (emotions, motivations) but of our "propriety, prudence and benevolence." He claimed that moral distinctions are not derived from reason but from sentiment which, for an Enlightenment philosopher, was a surprising turn. If we fast-forward to the 19th and 20th centuries, we can cite secular and religious thought from Marx, Freud, Neitzsche, and Bishop Fulton J. Sheen. The point is that serious thinkers have addressed human sociality without ever once mentioning neuroscience. Some still do.

In this chapter, I will address some of these approaches—some very recent—since they reflect on our social behavior from a perspective very different from mine. Why? Because it's useful, I think, to show the continuing breadth of interest in this subject. The idea that we have social impulses is deeply satisfying. That is, it's reassuring to have proof (or what passes for proof) that we are not 7 billion souls floating in a sea of troubles. People from an array of disciplines feel that they have the knowledge to show how we inexorably form connections. My three-pronged scientific approach, though different in methodology, is consistent with objective. I would also speculate, as I explain further during this discussion, that there are elements in these various approaches that can actually be integrated into my own, rigorous but broad approach.

Origins of Human Socialization
https://doi.org/10.1016/B978-0-323-85861-8.00005-0

Bowlby, plus

In recent decades, philosophical and economic approaches to sociality have shifted towards the psychological. In this vein, the post-Freudian British psychoanalyst John Bowlby (1969, and see Chapter 2) described what he termed "attachment theory." He sought to use evolutionary thinking, and ethological observations such as mother/young imprinting, to understand the primacy in human social development of the mother/infant bond. For decades, now, psychoanalysts and psychologists have referred to attachment theory as a significant formulation of the importance of the positively-valenced social bond (mother and infant) as opposed to bonds created by means of pecuniary exchange.

Nancy Chodorow, the prominent feminist psychoanalyst, follows Bowlby in *The Reproduction of Mothering* (1978), where she analyzes the mothering side of the baby/mother dyad that I discussed in Chapter 2. She suggests that while "the actual physical and biological requirements of childbearing and child care have decreased, mothering is still performed in the family, and women's mothering role has gained psychological and ideological significance…" (p. 4). In connection with showing that women have had and still have primary responsibility for the baby, she argues that they are as well the strongest influence on its development. She cites instinctual forces that motivate the mother to care for the infant, a "universal" which shifts the focus away from mothers' work and permits some theorists to "use biological claims to explain social forms" (p. 14). However, while Chodorow notes that such theorists are unable to provide "satisfactory evidence for a maternal instinct" (p. 18) in humans, it is now true that laboratory scientists working with lower animals have proven that rising estrogen levels in the blood, coupled with a complex role for progesterone (elevated levels followed by a precipitous drop of progesterone concentrations in the blood) predispose to maternal behaviors. So this raises the question of why, exactly, the mother's role in the mother/infant bond is so important to the baby's social development.

Chodorow (1999) separates from other prominent psychoanalysts and moves beyond drive theory, a simple reference to a primitive maternal instinct. In the alternative, she relies on (pp. 148–151) Object-Relations theory to provide a "rich and complex account of psychic functioning and agency that portrays a person in intrapsychic an interpersonal relationship." This theory argues, as does the present book, that people are fundamentally social. It emphasizes interpersonal relations primarily in the family and especially between mother and child. Thus, Chodorow provides an account of the mother's participation in the mother/child bond as it develops in real time and helps to encourage the baby's social development.

One of the more prominent American psychoanalysts to follow Bowlby was Daniel Stern. In *The Interpersonal World of the Infant* (1985), he divides early childhood into developmental stages which are unrelated to Freud's psychosexual stages but which, I suggest, involve the baby's comprehension of the social situation. Stern's concept of the baby's "emergent self" starts right after birth and concerns the baby's interoception (its awareness of the state of its body—hungry, cold, thirsty, etc.), probably with regard to the mother's supply of its physical needs. More than 2 months after birth, a sense of self (Stern's "core self") develops, representing the baby's side of the mother/baby dyad. Subsequently, a "subjective self" develops during which mother/baby social interactions become more complex and the baby's behavior extends beyond mere imitation. After a year, Stern suggests that the baby has a "verbal self," during which the baby can have more sophisticated social relations with the world. I find this analysis idiosyncratic, though in view of all the information that is constantly emerging, some aspect of it may yet turn out to have some relation to the truth.

Psychoanalysts are not the only nonscientists who have opinions on human sociality. The Introduction offers several examples. Anthropologists frequently deal with the prototypical social behavior, sex. Gray and Garcia (2013) assume an evolutionary perspective and emphasize a key practice associated with sex: reproductive access to females. In their opinion, male-male competition for females and the complementary function, females' choices among the competing males, take center stage. In these respects, various human cultures differ. Cultures which contain "despotic emperors, harems and eunuchs" (p. 38), powerful men with many wives, and polygynous patterns of sexual relations are necessarily different from cultures with polyandrous patterns (female with many males). Even cultures with pair-bonds and monogamous marriage differ. Some require arranged marriages (p. 67) while most do not. Physical features (e.g., weight) that are considered desirable vary from culture to culture as well (p. 71). The developmental path from childhood play towards sexual relations is still another cultural variant (p. 125). Some cultures ritually initiate boys and girls into sex at a relatively young age, e.g., just at the onset of puberty.

Thus, it is perfectly possible to propose a theory of human sociality that does not involve brain science. My argument is that it will not be possible, relying on such an approach, to propose a *comprehensive* theory, one which points to a full range of therapies if needed. As I have acknowledged, some of the material discussed above could, with some tweaks and elaboration, fit into one of the three strands of my own theory. But insofar as such material fails to broach neuroscience—the brain mechanisms for controlling everything from vision to hearing to a baby's instinct to reach out—it is simply not fully capable of describing what we now know about why humans are socially inclined. To be fair, I believe that psychology, psychoanalysis, anthropology, and other disciplines that study human actions are excellent at describing how individuals and even large groups understand the world. But what they cannot do is get inside the neurophysiology that has evolved over eons and made us *Homo sapiens.*

Biological anthropologists (who study diet, comparative anatomy, teeth, etc.) come closer to addressing how we evolved than do other members of their discipline, but they still do not get deeply involved with the genetic and hormonal data necessary to analyze exactly how sociality works (or worked a million years ago). In this book, I advocate using all the data at our disposal—anthropological, developmental/brain—and addressing it with scientific rigor. Psychological speculation is interesting—and can even make you feel good—but it won't get us to a deep (let alone complete) understanding of the human social impulse.

In the next section, I want to carry this discussion into the further reaches of speculation, since such speculation shows how non-scientific explanations of sociality can make sense (at least on their own terms) even while coming nowhere near explaining *why* we are social. A theory can make sense insofar as it explains specific social actions among some individuals. But that does not mean that it accounts for why, as a species, we are social and how we evolved to be that way.

Something completely different

The 14th century Tunisian philosopher and legal scholar, Ibn Khaldun, is described by his biographer Robert Irwin (2018) as "the greatest intellectual ever to have appeared in the Arab world." Thought probably to be a Sufi (p. 110), Khaldun's austerity and his occasional devotion to the Sufi practice of *khalwa*, "a temporary withdrawal from society in order to meditate" (p. 40), might not recommend him to hold forth on issues of social motivation. But some of his influential writing pursued the case where "in the harsh conditions of desert life, tribal groups of necessity develop a special kind of group solidarity which Ibn Khaldun called *asabiyya*" (p. 45). This word was defined as "a strong attachment, which holds several persons together to defend one another" (p. 45). Thus, for Khaldun, the instinct towards social affiliation is enforced by need, under the harsh conditions of desert life.

Back into the 21st century, the popular writer Tali Sharot, professor of Experimental Psychology at University College London, talks about a much different aspect of social life:a person talking to a group. In *The Influential Mind* (2017), she emphasizes the role of emotion in synchronizing one person's reactions with another. She asked couples how they met, and describes people "who could feel a connection immediately" (p. 42). She states that they "were inclined to attribute this sensation to magic." But in fact, it was what everybody else calls Chemistry. When one person sees that the other is attracted, he or she feels sufficiently self-assured to display their own feelings. The result is instant relaxation, one's guard is let down, and there is a sense of mutual recognition—"Oh, I felt like I knew him all my life." Undeniably, there was a connection.

What both of these accounts have in common is that they are limited to very specific circumstances. In the latter case, a lot of coincidence is involved. They are interesting a far as they go, but they are hardly scalable into a theory that describes where the human social impulse comes from. We could all, probably, offer instances from own lives that might be counterparts to these anecdotal claims. But they are just anecdotal, i.e., limited outcroppings along the path to what must be a much more comprehensive approach to sociality.

However, I would add that while anecdotal, psychological, and social science approaches to human sociality are inadequate, neuroscience on its own does not have all the answers. That is why I have adopted a three-prong approach that includes data from other disciplines. Only an appropriate, integral approach will be sufficiently nuanced and able to account for all of the relevant factors.

I will now discuss some innovative uses of technology in examining factors that influence human sociality.

Human social consciousness

During the last 20 years or so, neuroscientists have used magnetic resonance imaging (MRI, depends on oxygenated blood flow) with increasing facility to study the human brain. An important implication of this series of technical developments is that my colleagues can invent sophisticated experiments designed to test mental and behavioral abilities of subjects while in the MRI machine. *Changes* in blood flow correlated with changes in neuronal activity are measured in *functional* MRI (fMRI). The creative use of fMRI has led to new analyses of complicated social functions including, for example, altruism and "theory of mind" (see Chapter 8).

A leader in this field of social neuroscience is Matthew D. Lieberman, professor of psychology at UCLA. Recently, Lieberman's team was able to incorporate into his thinking about social behavior the so-called "default mode network" (DMN) of Marcus Raichle, neurologist at Washington University Medical School. The powerful DMN of the human cerebral cortex is activated when there is no immediate task at hand and the brain is at wakeful rest. It turns out that "DMN brain regions significantly increased connectivity during rest after encoding new social information (relative to baseline and post nonsocial encoding rest periods). Moreover, greater connectivity between the right side temporal-parietal junction (Fig. 1) and medial prefrontal cortex (MPFC), as well as other portions of the default network (vMPFC, anterior temporal lobe, and middle temporal gyrus) during post social encoding rest corresponded with superior social recognition and social associative memory. The tendency to engage MPFC and temperoparietal junction (TPJ) during rest may tune people towards social learning" (Meyer et al., 2019). Further, "hyperactivation to emotional

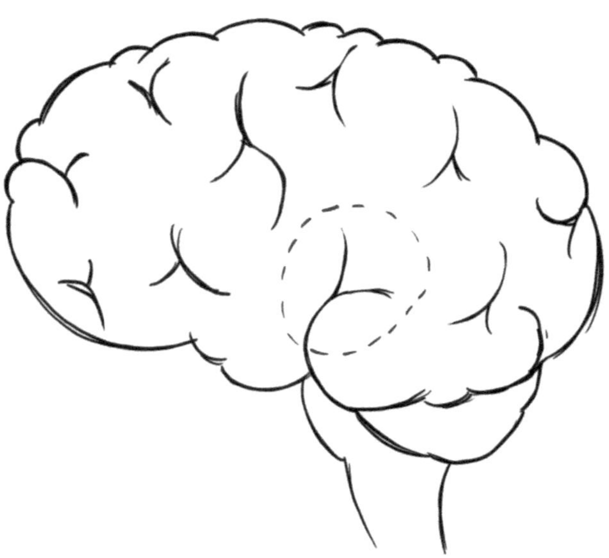

Fig. 1 The junction between the temporal lobe and parietal lobe of the human brain, a brain region clearly involved in regulation of human social behavior.

stimuli differentiated PTSD patients from controls, correlated with symptom severity, and predicted training outcomes." Critically, these effects were nonsignificant or marginal for non-emotional stimuli. Results were generally consistent throughout DMN and MNS. Unexpectedly, effects were non-significant in core affect regions, but robust in regions that overlap with the dorsal attention, ventral attention, and frontoparietal control networks (Tan et al., 2019).

In *Social: Why our brains are wired to connect* (2013), Lieberman argues that our brains respond to social pain and pleasure similarly to how they respond to physical pain and pleasure. "Social pain is a kind of real pain" (p. 45). Lieberman and his wife Naomi Eisenberger arranged for study participants to experience social rejection while being brain-scanned in the fMRI machine. The dorsal anterior cingulate cortex lit up. The authors knew that this brain region plays a role in the experience of physical pain and, in fact, has a high density of opioid receptors which would be instrumental in reducing pain. The point is that social pain, e.g., bullying, humiliation, or rejection, can be *like* physical pain in their intensity and tendency to leave sufferers with the understanding that they have been seriously harmed. If we know this, we will take mental (as well as physical) assaults seriously, and adopt aggressive measures to mitigate and prevent them. Mental assaults have frequently been dismissed with the claim "Oh, he was just joking." But if the pain that they cause is as real as other types of pain, then such dismissals (which are themselves a source of pain) should not be tolerated.

Lieberman has expanded his thinking about the social brain to issues of society's well-being (p. 241, ff.). He cites a 1995 study showing that while average income in America went up from 1946 to 1989, the curve for social well-being remained flat. He does not propose his own answers, but cites others' ostensibly promising initiatives, for example David Rock's Neuroleadership Institute and the need to educate junior high school students regarding their social motivations (p. 292). I must say, however, that whenever I see the prefix neuro- in front of anything but science, I am skeptical. When I looked up the Institute, I thought perhaps that Lieberman should have stuck to his fMRI results without recommending neurocorporate approaches. The Institute's Guiding Hypothesis is: "Organizational performance can be significantly improved, and the brain can show us how. By following the science, it is possible to develop significantly more effective talent strategies, and drive behavior changes across the globe...." This sounds at once facile and grandiose and, I think, could be misleading (let alone a misapplication of neuroscience to ill-defined business objectives).

It is an example of how rigorous neuroscience can be coopted and popularized largely beyond recognition. We all want to promote prosocial behavior, but not with unreal fixes.

In any case, the fMRI currently in use has shortcomings, mainly with regard to spatial and temporal resolution and excitation/inhibition. As to spatial resolution, we know that neural activity in the brain is organized on a cell-by-cell basis. Two neurons within a small fraction of a millimeter from each other could have activities with totally different meaning for a particular behavioral function. The fMRI, however, does not have the spatial resolution capacity to track that phenomenon. As to temporal resolution, while the dynamics of a nerve cell's response to some stimulus are measured in milliseconds, the fMRI is not fast enough to keep up. Worse, when there is a spot of blood flow/neural activity, the fMRI can't tell whether those neurons are excitatory, inhibitory or both. Nonetheless, the creative use of fMRI—even in social neuroscience—is likely to play at least some part in social neuroscience for the foreseeable future.

The field of social neuroscience has, in fact, become enormously popular in the last several years, as theorists test the limits of how our brains are designed for social connectivity. These are outward-facing theorists who, unlike the psychoanalysts, are concerned with the individual's connection to the world-at-large, rather than with his or her role in a specific (usually dyadic) relationship (e.g., the bond between mother and child). I see this type of psychology (and engineering) as a start towards reorienting the conversation in neuroscience towards the idea that there are many arguments—many of them complementary—that investigate how humans are instinctively social.

It is important, now, to attempt to meld purely psychological thinking with our first inroads into neural mechanisms in humans. In *Consciousness and the Social Brain* (2013), Princeton University psychology professor Michael Graziano seeks to meld studies of visual attention and perception with the handling of information by the brain to create a new theory of social consciousness (by "consciousness," I mean the state of being awake and aware of one's surroundings, i.e., the mind's awareness of itself and the world). He calls this *attention schema theory*, in which a person's consciousness subsumes information of which the person is aware into the act of awareness itself (pp. 11–21). Visual perceptions—say, the image of a person—become descriptions (and possibly identification) of the person. The perception exists as social information. With that visual perceptual information in place, the perceiver must be aware of that information. Accordingly, if we are more in sync with our environment because the very act of perception draws us into the environment, then our motivations can no longer be self-involved. When what we perceive is a person, our perception will predispose us to a prosocial act.

This is an extraordinary idea. It means that the *state* of being aware—which is continuous—translates into constant social awareness, not just isolated acts of becoming aware. Further, it means that insofar as we are aware of another person, we will necessarily and continuously have a reason to reach out to that person: we are in their space and they are in ours so long as we are conscious. According to Graziano, social consciousness is part of being alive; it might well be considered instinctive.

Of course, awareness depends on prior activation of brainstem mechanisms supporting arousal (Pfaff, 2019), but for Graziano there is more. In his view, awareness is computed from perceptual categories, chunks of information bound together. It depends on attention, "an enhanced way of reacting to incoming stimuli" (p. 23). Attention depends on top-down influences which I referred to for the visual system (Chapter 5) and the auditory system (Chapter 6). Those top-down influences result from "biases competition" which, I presume, are emotion- and motivation-dependent. "Attention is a data handling method." (p. 25)

Moreover, *social* consciousness must depend on the processing of social stimuli (in this book, Chapters 5 and 6). In those chapters, I highlighted face perception areas, for example in a subdivision of the parietal cortex and the inferotemporal cortex (Fig. 1; Chapter 8). Awareness functions applied to such social signals make for social consciousness.

Then, to use a baseball analogy, Graziano swings for the fences. In his thinking "consciousness is the result of social intelligence, our ability to understand the minds of other people, turned inward on ourselves." (p. 134). While recognizing previous statements of this general idea, Graziano feels that his attention scheme theory benefits from its specificity: How do we understand the attentional state of another person? In this respect Graziano's thinking benefits from the tremendous amount of work that went into Theory of Mind neurophysiology (critiqued in Chapter 8)

Further, Graziano's theory does not lead to a *dei ex machina*, a magic agent that pseudo-solves a problem. Although the theory refers to specific neuronal regions (see above), it does not confuse the simple naming of those brain regions with a functional explanation.

Graziano carefully distinguishes his theory from University of Wisconsin professor Giulio Tononi's popular "integrated information theory," for three reasons (pp. 139–149). First, integrated information is not always consciousness. Second, how would an experimentalist test it? Third, Tononi's theory does not explain how the information is reported. How is the information, as integrated, explicitly acknowledged? I share Graziano's reluctance to accept Tononi's theory (see Pfaff, 2019) for two additional reasons. Tononi's theory assumes its own conclusions and, while doing so, requires the reader to wade through unnecessary mathematical constructs.

Social neuroscience has attracted many theorists, not all of them so well-focused as Graziano. Jay Evans Harris' *Minding the Social Brain* (2013) covers a lot of territory—from Freud to Gerald Edelman's theory of neuronal group selection to current work in social neuroscience. While Harris' book, like much of psychoanalysis, suffers from overreach, it does make a few points useful to neuroscientists. For example, Harris suggests that when the anterior cingulate cortex picks up signals of social dissonance, we feel "superego anxiety" (p. 232). We notice this when we attend to signals from our interior body, our guts. The anterior cingulate cortex is part of Marcus Raichle's "default mode network" (widespread and active in the absence of any explicit task) (Raichle and Snyder, 2007; Raichle, 2010). Harris theorizes that this connectivity "uses social generalizations [to tell us] what changes are necessary in our social approach or social avoidance so as to generate our best survival possibilities." (p. 238).

Neuropsychoanalysis

Mark Solms, a South African psychoanalyst with bases in London and New York has led a comprehensive attempt to merge psychoanalysis with modern neuroscience. As Solms and psychologist Jaak Paanksepp (Panksepp and Solms, 2019) wrote, "Neuropsychoanalysis seeks to understand the human mind, especially as it relates to first-person experience". It recognizes the essential role of neuroscience in such quests. However, unlike most branches of neuroscience, it positions mind and brain on an equal footing. "Neuropsychoanalysis accordingly counters prevailing extreme reductionism in neuroscience and biological psychiatry." Neuropsychoanalysis, by recognizing the deep evolutionary roots of the human mind and emotional disorders, seeks a more coherent understanding of the primary-process brain affective networks of mammalian brains than currently exists in psychology. As a side point, I was grateful that neuropsychoanalysts were quick to recognize the large number of genetic, biochemical, endocrine and neural functions shared between animal brains controlling sexual motivation and sex drive (libido) in humans (Pfaff, 1999).

More generally, Panksepp and Solms (2019) listed “key contributions of this fledgling approach to the mind”: (i) the full recognition that the foundations of human emotions can be clarified in animal models, and that the neural constitution of emotional feelings can finally be empirically elucidated, (ii) "identification of the motivational and emotional substrates of the denial of left hemiplegias (weakness on one side of the body) in neuropsychological patients with anosognosia (unawareness of one's own illness), (iii) clarification of the fact that dreaming and rapid eye movement (REM) sleep are dissociable, with dreaming being strongly energized by midbrain dopamine dynamics, via the general-purpose reward seeking,” and (iv) recognition of emotional feelings in animals has wide implications for understanding psychiatric disorders. For instance, neuropsychoanalysis offers novel cross-species perspectives on many emotional disorders, such as addictions and depression.

Years later, Solms (2018) extended his view of what neuropsychoanalysis can do by making certain core claims concerning “the emotional mind”: First, the human infant is born with innate needs, which can only be met through social connection; second, through ego development we learn to meet these needs—sexual and otherwise—in the world we face. Further, as a therapy, psychoanalysis aims to improve “deeply automatized action plans”, which are dependent on what Solms identifies as “dominant emotions” which are felt but not consciously identified as such. Finally, he presses the point that “psychoanalytic therapy achieves good outcomes—at least as good as, and in some respects better than, other evidence-based treatments in psychiatry today,” and that psychoanalytic psychotherapy is equally effective as other forms of evidence-based psychotherapy.

Perspective

This chapter represents a very small portion of what nonscientific intellectuals have said concerning human social life. Indeed, there are whole publishing imprints, like HarperOne, devoted to popular musings on this question. So, my point in examining this work is not just to criticize it—though there is much to criticize—but to demonstrate that there is both an enormous awareness of our social nature and a desire to explain it. In seeking to explain why we are social we explain not just ourselves as individuals but the nature of the social world. There is a certain appeal in naturally extrapolating from the individual to all of humanity as if everything were subject to one overriding principle uniting us all. As secularism replaces religion, such claims have great, deeply satisfying appeal. If you can think of yourself as naturally knit together with others—by force of your mental make-up—you feel like you are part of something larger than yourself. This is what religion does. To a degree, we are replacing religion with pop-psychology, information theory, anthropology, and science-lite. The world of ideas is morphing, quickly.

But still. If we view all these non-scientific claims in toto, they add up to a multiplicity of approaches to human connectivity that, in a way, is parallel to the approach that I am suggesting in this book. There *has* to be more than one discipline involved in determining why we are social and how we got that way. What I am trying to do is to reduce the cacophony of voices offering an explanation to a few inter-related, scientific arguments. Hobbes and Rousseau, whom I cite in the Introduction, would probably have denied any attempt to replace religion with a political argument for how society came to be. But they (and others) initiated an effort to come up with some plausible real-world rationales... which now you see repeated in some of the efforts that I have reviewed. There is a long line of such efforts that preceded hard science, and these efforts will continue because everyone is interested in the subject. My contribution, relative to these efforts, is to redirect them towards science where possible.

Accordingly, I set out a three-pronged approach to sociality that draws on different sciences (developmental psychology, neuroscience) and a basic concept in anthropology (evolution) to suggest how the human social impulse originated. I am not embarrassed to say that science aspires to push softer approaches aside by offering empirical evidence. Science is just as interested in the Big Picture and in grand, impressive theories as any other intellectual program (or even religion). But it believes in arriving at those large ideas through individuated, verifiable steps that can be reproduced (or falsified) by disinterested observers. Some of the approaches that I have discussed in this chapter follow that *modus operandi*. In fact, they demonstrate why partial approaches to objective answers make science that much more crucial if we are to understand basic issues of human nature. That we are profoundly and irrevocably social beings *is* a big, overweening idea, but my goal has been to take you through some of the nitty-gritty underneath it to show you how science got there (okay, half a billion years after the fishes did).

In the next chapter, I discuss altruism, a direct expression of the social impulse.

New questions

1. For the psychoanalyst in the age of novel coronavirus-19, what to tell her patient? As warned in my introduction, patients probably feel conflicts between their natural social instincts—Which this book explains in three intertwined ways—And governments' warnings with regard to "social distancing." What to do?

References

Bowlby, J., 1969. Attachment and Loss. Basic Books, New York.
Chodorow, N.J., 1978. The Reproduction of Mothering. University of California Press, Berkeley, CA.
Chodorow, N.J., 1999. The Power of Feelings. Yale University Press, New Haven, CT.
Gray, P.B., Garcia, J.R., 2013. Evolution and Human Sexual Behavior. Harvard University Press, Cambridge.
Graziano, M.S.A., 2013. Consciousness and the Social Brain. Oxford University Press, Oxford.
Harris, J.E., 2013. Minding the Social Brain. International Psychoanalytic Books, New York.
Irwin, R., 2018. Ibn Khaldun: An Intellectual Biography. Princeton University Press, Princeton, NJ.
Lieberman, M.D., 2013. Social: Why our Brains Are Wired to Connect. Crown, New York.
Meyer, M.L., et al., 2019. Evidence that default network connectivity during rest consolidates social information. Cereb. Cortex 29 (5), 1910–1920.
Panksepp, J., Solms, M., 2019. What is neuropsychoanalysis? Clinically relevant studies of the minded brain. Trends Cogn. Sci. 16, 6–8.
Pfaff, D.W., 1999. Drive. The MIT Press, Cambridge.
Pfaff, D.W., 2019. How Brain Arousal Mechanisms Work. Cambridge University Press, Cambridge.
Raichle, M.E., 2010. Two views of brain function. Trends Cogn. Sci. 14, 180–190.
Raichle, M.E., Snyder, A.Z., 2007. A default mode of brain function. NeuroImage 37, 1083–1090.
Solms, M., 2018. The scientific standing of psychoanalysis. BJPsych. Int. 15 (1), 5–8.
Stern, D., 1985. The Interpersonal World of the Infant. Basic Books, New York, NY.
Tan, K.M., et al., 2019. Posttraumatic stress disorder and the social brain: affect-related disruption of the default and mirror networks. Depress. Anxiety 36 (11), 1058–1071.

Further reading

Raphael, D.D., Macfie, A.L., 1976. On Adam Smith, the Theory of Moral Sentiments. Liberty Fund, Carmel, IN.
Sharot, T., 2018. The Influential Mind. Little Brown, Boston, MA.

CHAPTER 8

Social behavior at its best: Altruism

Yesterday, it happened again (Fig. 2 in Introduction). A man leapt onto the track to save another man who was stuck there, "staring ahead absently." As an example of such positive human social behavior, my New York ideal from years ago is Wesley Aubrey, a Black construction worker who left his own child on a subway platform to jump on the tracks, cover a trapped epileptic, a young white man, with his own body, and save him from an oncoming train. Afterwards and to great fanfare, Aubrey said that he never gave his act a second thought—it just came naturally. He was no neuroscientist, but he was right. Altruism, the willingness to give without any expectation of reward, is an extreme but also utterly commonplace human social impulse. We do not give because we think it's "right," but because we do not think. We just do it. As I will show, our brains are designed so that—should the need arise—we put ourselves in other people's places and act empathetically.

Building on early work from Yale University's Karen Wynn (Chapter 2), Rodolfo Cortes Barragan and his lab head, Andrew Meltzoff (2020) at the University of Washington, report that "19-month-old human infants repeatedly and spontaneously transferred high-value, nutritious natural food to a stranger (Experiment 1) and more critically, did so after an experimental manipulation that imposed a feeding delay (Experiment 2), which increased their own motivation to eat the food." Hungry kids would normally grasp at food, so such generosity—at such a young age—suggests a natural inclination to share irrespective of the personal consequences.

The data from Meltzoff's lab are consistent with the report by Felix Warneken and Michael Tomasello of Leipzig's Max Planck Institute, referred to in Chapter 2, stating that "human children as young as 18 months of age (prelinguistic or just-linguistic) quite readily help others to achieve their goals in a variety of different situations. This requires both an understanding of others' goals and an altruistic motivation to help." In this regard, the Meltzoff work used a clever experimental setup in which the "experimenter would pick up and fumble each of a bunch of attractive, tasty fruits,

Origins of Human Socialization
https://doi.org/10.1016/B978-0-323-85861-8.00007-4

accidently dropping them onto a tray. The experimenter tried to reach for the dropped fruit with his outstretched arm (seemingly begging for the food), but he could not attain the food, because he was blocked." Almost 60% of the infants confronted with a "Begging Experimenter" picked up a fruit and gave it to the adult, and only 4% did so when the experimenter was not begging.

Then Meltzoff made it even tougher for the baby to offer the food to the Experimenter. The babies came to the lab at a time just before their expected meal or snack, so they were hungry. They were really hungry because, after the crucial behavioral test, they ate more. Nevertheless, almost 40% of the infants gave fruit to the "Begging Experimenter," compared to 0% in the "non-begging" control. So, we know that babies considerably less than 2 years of age, without any kind of special training, will perform a purely altruistic act.

The crucial fact is that altruistic impulses do not stop with infancy—e.g., see Wesley Aubrey. Rather, altruistic behaviors follow naturally from human brain mechanisms. I have argued (Pfaff, 2015) that the elements of altruistic behaviors can be explained on the basis of well-understood brain mechanisms that allow us to visualize the behaviors in advance and literally identify with the person to be helped before we actually help them. This extraordinary mechanism brings us into a type of mental contact with that person, making our action seem plausible, feasible, and something we could imagine for ourselves-in-need. That is, empathy—a sense of shared feeling—is wired into our brains. We do not necessarily learn empathy from religion or first-grade teachers. Rather, we are instinctively empathetic. The Golden Rule ("Do unto others....") is not a command so much as it is a universal formulation of how we were designed to behave—i.e., how our brains were designed to instruct us to behave. Moral systems all over the world, for centuries, have some version of the Rule—not so much to enforce it but to acknowledge its acceptance.

Accordingly, any discussion of human brain mechanisms that favor social behavior would be incomplete without examining our natural altruism. Just as smiling at someone we see or hear is a neurophysiological responses to external stimuli, so is altruism. Specific regions in the brain motivate us to offer a piece of food or jump on top of someone when a train is hurtling toward them. There is no moral imperative in a greeting, and there is none in a selfless, altruistic act. Nike's trademarked motto, "Just Do It" (which has its own Wikipedia entry because it's so universal), could just as easily be applied to the brain of a person confronted by someone in need. The fact is that we are not designed to be bystanders but, rather, active participants in each other's survival. As I have argued throughout this book, evolution is not concerned with anyone in particular but is absolutely driven by whatever promotes the species' survival. Altruism operates at that level, starting in our brains.

Consider these steps. First, *our behaviors are represented to our nervous systems before they occur.* Early on, this neurophysiological fact was given the German name "reafferenz" mechanisms. A more illuminating name is "corollary discharge": milliseconds before a motor act is initiated, the relevant sensory systems are informed. This set of mechanisms helps account for the stability of the perceived world. Think about it: when you move your eye to the right, does the world appear to move to the left? After all, it should appear so, as a consequence of the image on the retina. But it does not because the visual system is primed for the eye movement to the right, by corollary discharge to the visual system from the oculomotor system.

Second, *perception of the individual toward whom the benefactor is about to act.* This is easiest to think about when we involve the visual system, carrying the image of the individual and was treated here in Chapter 5. We have known for more than 30 years that our brains contain neurons which respond to faces. Now we also know that some neurons in the prefrontal cortex respond to the vision of social interactions.

Third, *merge or "blur" the images of the person acting with the image of the person toward whom the benefactor will act.* This is the most crucial step for altruistic behavior and is also the easiest. Think about it. The images are blurred in our perceptual neural pathways. Is it easier for the visual system to perform less precisely (as in merging or blurring images) than to require the highest degree of precision? Even though visual pathways and cortical mechanisms for vision are beautifully engineered to provide precise imaging, altruistic behavior allows for a relaxation of those mechanisms, to allow the individual to behave toward the other as (s)he would behave toward him- or herself (Pfaff, 2015).

Fourth, *following obviously from the third step,* if the behavior previewed is beneficial, kind, helpful, then the answer is yes, do it. If it is harmful, don't.

Certainly, these elementary mechanisms will be understood in greater detail as neuroscience moves ahead. Further details and additional considerations will emerge. For example, Leonardo Christov-Moore and his team (2020) have written that "recent task fMRI studies suggest that individual differences in trait empathy and empathic concern are mediated by individual differences in patterns of connectivity between self-other resonance and top-down control networks that are stable across task demands. Using machine learning, we demonstrate that patterns of *resting state fMRI connectivity* of resonance and control networks predict trait empathic concern." That's complicated, but the work marries current neurophysiological work with our questions about empathy. This is a specific example of what my theory, noted above, says: natural, other-directed (empathic, altruistic) feelings mediated by normal brain networks.

Another example, explored with a different state of the art technique, optogenetics, developed in the famous Stanford University lab of Karl Deisseroth. In this technique they used viral DNA which, on the one hand, can be expressed in a particular type of nerve cell, and, on the other hand, regulates nerve cell excitability. Using this technique, Jennings et al. (2019) discovered neurons in the frontal cortex which respond selectively to social stimuli. These almost certainly include neurons which would be involved in altruistic behaviors toward others

Thus, we are beginning (at least to some extent) to visualize and reproduce the actions of altruistic brain mechanisms by actually looking at the brain and manipulating nerve cell excitability. *We can see altruism in action at the level of specific regions of the brain.* This is analogous to what neurobiologists can do with vision and hearing. While this may sound unnecessarily reductive with regard to so complicated a movement as saving someone, the fact is that altruism, like vision and hearing, originates with specific operations in the brain. We may think of ourselves as a moral species exempt from "nature red in tooth and claw," but actions that we call moral actually rely on neuronal activity in specific areas in the brain.

I concede that this amoral or, rather, non-moral explanation for what might be called a generosity of spirit is likely to be disconcerting. It feels somehow reductive, in that it extricates kindness and generosity from everything we have learned about how to behave. It makes much of the Sermon on the Mount sound like yesterday's news. It deprives us of our "better nature," replacing it with neurophysiology over which we have little choice. Of course, the push-back is: who even *wants* to be robotically decent? We'd rather say that through hard work, diligent Sunday School attendance, and there-but-for-the-grace-of God attention to the *Times*' Neediest Cases Fund we purposefully got to be kind. Okay. But think of it this way. If, on the one hand, we don't need external rules to have prosocial impulses, isn't that reassuring? Doesn't it make human nature that much more admirable? If we can appreciate that human beings naturally evolved to be altruistic, we are in a better position to respect ourselves and each other. Where the Christian religion presumes that we are "fallen"—and would discipline us accordingly—brain science proves just the opposite. It offers some reason for large-scale hope.

Indeed, every day during the coronavirus pandemic, the news was full of stories of front-line medical personnel working 18-h shifts, personnel who by doing their heroic work exposed themselves to infection. When protective gear was in short supply, they went in anyway. When they stopped long enough to be interviewed, they invariably said "That's my job," the moral equivalent of "That's who I am." They could *not* imagine *not* giving of themselves. One doctor, asked about how she was doing outside of work, said "That's all there is." The pandemic offers a kaleidoscope of case studies in reflexive, unmediated altruism. I suggest that most of us in the position of these workers would act no differently. While the media celebrated them as heroes, they staunchly denied it. They responded to a human need because they saw one.

Corruption. But don't get carried away with self-congratulation. Altruism has a dark side. That is, when altruistic brain mechanisms work, they can produce results which (ironically) do not benefit society as a whole. Instead, the benefit accrues only to a small local group out to benefit themselves. In this regard, consider corruption, the "I'll scratch your back, you scratch mine" relationships. We see this sort of behavior in every area of life where there's a buck to be made or illicit favors to be exchanged.

A small sketch suffices. Consider the policeman walking the beat in a tough neighborhood. He's hungry. His beat includes the small delicatessen owner, who wants to park his car in a no-parking space right in front of his deli. Their brains, operating normally, arrange for the deli owner to offer free donuts and coffee to the policeman, who then overlooks the parking violations. It all seems so natural. We see the same sort of action in Russia, famous for rampant corruption, and among the old-style pols of Chicago (jailed Illinois governor Rod Blagojevich being just an updated version of operatives in a city dubbed "the most corrupt" in the country).

Our altruistic brain mechanisms can be thought of as "neutral," almost always producing good behavior, but sometimes producing corrupt behavior defined as "bad." It is the neutrality that's interesting, in that morality never enters into the operation of the mechanisms' basic operation. Their operation is situational. Each party sees a need and acts to meet it. While ostensibly good altruism never stops to calculate whether there is a reward, in socially "bad" altruism the reward is subsumed in mutual altruism. In corruption, everything works just fine to produce a result that—from an external, moral perspective—is undesirable. When this occurs, the brain has not weighed externally imposed standards.

So, if insight into brain mechanisms casts new light onto our altruistic behaviors, how did these mechanisms evolve? We know that these mechanisms evolved to promote our species' survival, but is there evidence in the evolutionary record suggesting how they arose?

Evolutionary thinking

Long before I began thinking about brain mechanisms underlying altruistic behaviors, evolutionary biologists were studying these mechanisms and theorizing their role in human and animal societies. Theodosis Dobzhansky, a famous geneticist, observed: "Nothing in biology makes sense except in the light of evolution." So, I want to consider how evolutionary thought adds to our understanding of altruistic behavior.

I propose that three evolutionary theories that have emerged independently and may, at times, even be at odds, can actually be considered in tandem—much as in my braided theory of social behavior—to provide a backdrop to the altruism that we observe in *Homo sapiens*. These three theories are "Selfish DNA," "Kin Selection," and "Group Selection."

Basically, selfish DNA theory states: "I behave with self-interest to make sure my own DNA is passed on to the next generation." As the British biologist Richard Dawkins suggests in *The Selfish Gene,* each of us behaves so as to increase the frequency of our genes in the gene pool. Each of us whose behavior is affected by a given gene will seek to survive as long as possible and reproduce as frequently as possible. This is a hard-nosed perspective. Of course, this starts out sounding like the opposite of altruistic brain biology, but it will get subsumed into the two subsequent steps, as follows.

Kin selection theory states: "I help this person because he is related to me and if he survives through the age of reproduction, parts of my DNA will be passed on. The closer he is related, the more of my DNA he has." The originator of sociobiology, Edward O. Wilson (*The Superorganism*), ascribes the theory to the British biologist J.B.S Haldane in the 1930's. Cambridge scientist William D. Hamilton gave this theory an arithmetic caste: Multiply the benefit of your impending altruistic act by the closeness of your relation to the beneficiary of your act. If that multiplication product exceeds the cost to you, then do it.

And then, group selection theory states: "I help this person because he is part of the same group I belong to. If the group survives, then both my kin and I can pass on our DNA to the next generation." Harvard University biomathematician Martin Nowak believes that within a group "cooperation can emerge out of nothing more than the rational calculation of self-interest." Group selection and kin selection both enhance the instinct toward altruistic behavior, but they do so fractionally, through implicit calculation.

Geneticists and biologists have argued about the relative potency of these theories to account for the evolution of non-human primate and human altruistic behaviors. But I think that all three of the evolutionary mechanisms that they describe could have operated simultaneously. "Selfish DNA" is the narrowest and most ruthlessly efficient, but kin selection builds on it and group selection can be thought of as an expansion of kin selection. My conjecture is that the three theories do not negate each other. Rather, they augment each other, with their relative importance varying according to the time period and culture in question.

Evolutionary thinking about altruism has also profited from studying the behavior of nonhuman primates. Frans de Waal first became famous because of his discoveries about the behavioral biology of the bonobos (1997). As opposed to some popular ideas about the selfishness of certain primates, deWaal showed bonobos to be prosocial, thoughtful, empathetic. For example, one bonobo worried about getting too much food relative to others. As a reward for success in cognitive tests, she earned milk and raisins while other bonobos looked on. But "after a while she refused all rewards. Looking at the experimenter she kept gesturing to the others until they, too, got some of the goodies. Only then did she finish hers" (p. 190, De Waal, 2009). The behavior is particularly interesting because this bonobo earned her food fair and square and *still* required its redistribution. It's not as if she reacted against blatant favoritism. Her response was on a continuum with self-made billionaires who have taken a pledge to give away most of money.

In his 2009 book, deWaal shows that chimpanzees can share, even under the same circumstance where they have a sense of ownership (p. 6) Yes, each chimp wants to be the first to put his hands on the incoming food, but by about 20 min, because of sharing, "all the chimpanzees in the group will have some food." As opposed to "macho origin myths" which emphasize the importance of individual strong males, we descend from a long line of group-living primates with a high degree of interdependence" (p. 21).

Nor is altruism limited to primates. At the University of Chicago, the neurophysiologist Peggy Mason (Bartal et al., 2011) used rats to prove that they can show empathic behavior. The experimenters "placed a free rat in an arena with a cagemate trapped in a restrainer. After several sessions, the free rat learned to intentionally and quickly open the restrainer and free the cagemate. Rats did not open empty or object-containing restrainers."

Furthermore, the rats shared chocolate. In subsequent experiments Mason's team sowed that empathic behavior could be altered by social experience. "Pair-housing with one rat of a different strain prompted rats to help strangers of that strain, evidence that rats expand pro-social motivation from one individual to phenotypically similar others. To test if genetic relatedness alone can motivate helping, rats were fostered from birth with another strain and were not exposed to their own strain. As adults, fostered rats helped strangers of the fostering strain but not rats of their own strain." (Bartal et al., 2014). These results with rats certainly support my theory that altruistic behavior is built into the brains, because rats certainly do not have the legal, religious or human cultural forces that might otherwise have been argued to have causative roles.

It probably will not surprise anyone that elephants are smart enough to show empathic behavior. DeWaal cites a healthy sighted elephant who took care of a blind elephant. "The blind one was clearly dependent on the other, who seemed to understand this" (p. 121).

Human evolutionary biology. But back to humans. Harvard evolutionary biologist Richard Wrangham (2019, p. 9) states that humans are low in *reactive aggression* ("hot," "losing our temperature and lashing out"), but high in *proactive aggression* ("cold," "planned and deliberate" aggression). Wrangham's picture maps perfectly onto a dissection of brain mechanisms of aggression published many years ago, as mentioned above, by John Flynn, a former Catholic priest who, as a Yale professor worked at the Connecticut Mental Health Center. Flynn electrically stimulated small groups of neurons in the hypothalamus of awake, well-behaving cats. Stimulation of the ventromedial hypothalamus led immediately to their enraged biting attacks on whatever was nearby—it was like Wrangham's "hot" aggression. In dramatic contrast, stimulation of the lateral hypothalamus caused a crouched, stalking behavior as though preceding a hunting attack—like Wrangham's "cold" aggression. I do not think it accidental that Wrangham's sociobiological thinking and Flynn's neurophysiological results coalesce in the same dichotomous account of aggressive behaviors.

Further, Wrangham offers an interesting explanation of how the domestication cited in Chapter 1 came about. Wrangham favors "genetic selection against reactive aggression" during which modern humans lost more of this aggressiveness from some common ancestor than Neanderthals did. In his unusual search for "the origins of humanity's peaceful virtues" (p. 168), he argues for "300,000 years of selection against reactive aggression", as required for our "evolutionary history of self-domestication" (p. 223). On

p. 49, he states "We are calm, compared with a typical wild animal." Tiny children do not strike out against each other. The willingness to interact peacefully with others, even strangers is inborn.

Eminent biologically trained thinkers chime in to reinforce some aspects of the neuroscience/altruism link argued in this chapter. Renowned Harvard evolutionary biologist Edward O. Wilson's (2019) book includes findings from a wide variety of species to remind us that there is safety in numbers, implying that altruistic behaviors within a group are certainly a good idea. And Adam Rutherford's (2019) work includes findings from a wide variety of species and to argue that "biology enables culture", thus reinforcing the general idea that altruistic behavior does not always spring necessarily from purely "non-biological" causes. These books are valuable because they bring in data from species that neuroscientists like myself could not maintain in our laboratories.

Thus, it is apparent from the work of evolutionary biologists that the human social impulse, expressed in terms of altruism, is traceable to our evolutionary past, both with regard to non-human primates (from whom we branched off on the evolutionary tree) and early humans. We can even see the operation of altruism in non-primates, such as rats. The point is that altruism evolved across species as a protective *modus operandi*. Obviously, no moral guidance was involved. The studied behaviors were instinctual, emanating from dedicated brain mechanisms. Altruism, as a behavior, exemplifies how the three prongs of my theory of human social behavior combine into one explanation: the dedicated *brain mechanisms evolved*, and are discernable in *babies' development*.

Theory of mind

Frequently, we are inclined to act altruistically even though the objects of our action have said nothing about their condition, let alone asked for our help (consider the man lying glassy-eyed on the subway tracks). In such circumstances, we infer the mental state of another person that we are moved to help. This ability to grasp another's mental state has an unfortunate name, "theory of mind," which implies that it is a theory of everything mental. The phrase, regrettably, is sometimes used in connection with our understanding of altruism.

In my field, we are pursuing two huge groups of initiatives to advance our understanding of how humans may be able to help another individual even when a person did not signal exactly what he or she needs or wants.

First, we design experiments using experimental animals to demonstrate that, yes, under some circumstances some animals have this ability. Second, for human subjects, we are trying to deconstruct "theory of mind" to infer all the component mental processes involved. These include perceiving others, especially their faces; inferring their emotional state; understanding causality (what behavioral acts lead to what results); and many other component processes. Of course, "identifying with others" as I described above, an essential step in altruistic behavior, is also a crucial brain function in behaviors associated with "theory of mind."

Sara Schaafsma and I cooperated with the team of CalTech professor Ralph Adolphs (2015) to criticize the "theory of mind" field, knowing that the meaning of the phrase is often vague and inconsistent, that its biological bases are, as a consequence obscure, and that the methods used to study it are impossibly heterogeneous. We argued for a reformulation of the phrase through a systematic two-stage approach. First, we have to finish the deconstruction of the "theory of mind" phrase into a comprehensive set of basic component processes. Second, when that is finished, we can start the complementary reconstruction after which a scientifically decent concept of how humans infer others' needs and intentions can be studied properly in human and animal brains.

In any case, what neuronal groups in the human brain are most important for "theory of mind" abilities? Some ask the oversimplified question "is there a theory of mind network?" I would say that at the present time, we don't really know. Admitting that, it is nevertheless fair to say that one cortical region has received a lot of attention: the junction between the human temporal cortex and parietal cortex (Fig. 1 in Chapter 7). Scientists in the Duke University laboratory of Scott Huettel (Carter et al., 2012) used functional magnetic resonance imaging (fMRI) to track brain activity during social conditions involving a simplified poker game. Bluffing was part of the game, so that the human participant being recorded had to decide whether to bet or to fold. The temporo-parietal junction was the area of the brain's activity that best predicted the subject's decision. This activity was particularly important when the opponent judged to be superior was human, rather than a computer.

Bitsch et al. (2018) added detail to such findings, while confirming the importance of the temporo-parietal junction. These authors used a prisoner's dilemma game (a form of "do I cooperate or compete?") in which "participants interacted with playing partners who behaved according to stable strategies either competitively, cooperatively or randomly." The temporo-parietal junction region on the right side of the brain (and the posterior-medial prefrontal cortex) "showed higher activity during the interaction with a competitive compared with a cooperative playing partner." Only the right temporo-parietal junction showed a high response *during an early interaction phase*, before the decisions in the game became unpleasant. So, these data added some neuroanatomical detail and also added some behavioral specificity. However, a lot remains to be discovered concerning the neurobiology of "theory of mind" behaviors.

For example, we can't avoid recurring to the mirror neurons that I described earlier in monkey-see-monkey-do experiments. In Chapter 5, I suggested that mirror neurons seem ideally suited to putting oneself in the position of another person and, consequently, empowering empathic/altruistic behaviors. Through these neurons, we literally go through neuronal activation for the same motions as the other person and can, at least theoretically, experience what they experience.

These capacities for empathic behavior did not just develop yesterday. From the patterns of wounds in ancient skeletons unearthed, we can guess that even Neanderthals took care of wounded members of their tribes. Was that behavior due to mirror neurons? We don't know and most likely never will know.

In any case, if we ever work out the kinks in "theory of mind," it could provide significant support for humans' capacity to empathize with other humans. We are not mind readers, but we pick up cues. Theory of mind could encompass humans' capacity to recognize and *read* faces, e.g., being able to tell when someone is happy or distressed. We can already read body language and know, for example, when someone is feeling faint and may need support. Theory of mind could easily segue into all the ways that we learn to understand what others need. As we begin to program Artificial Intelligence to understand emotions (see my *A Neuroscientist Looks at Robots*), and as we put our capacities into robots, we will obviously have to understand, in vastly more detail, how we do what we do. Ironically, the massive interest in AI and its inevitable advance is turning us back on ourselves. In this sense, robots are evolving from us.

Network multipliers

In this chapter, I have addressed individual behaviors, although all of us behave in social contexts that can be pictured as networks. Along these lines, a mathematically inclined sociologist at Yale, Nicholas Christakis (VanderWeele and Christakis, 2019), has used the "spillover effect" which is "sometimes used in the social and biomedical sciences to describe one person's exposure affecting the outcome of another." In other words, they are thinking in terms of the effect of social networks. Based on that idea, they came up with an equation using the concept of "a network multiplier (NM), defined as the ratio of the total number of changed outcomes due either to someone being directly treated or to spillover, as compared with the number of changed outcomes due only to someone being treated directly." That is, the effect of each person in the network on others in the network is multiplied by the network. Likewise, Christakis (Shakya et al., 2017) used "experimental and observational analyses to map face-to-face social networks in Western Honduras, and then conduct a randomised controlled trial of a friendship-based network-targeting algorithm" to exploit his understanding of social network structure (Wu et al., 2018). Further, Christakis and his collaborators (Shirado et al., 2019) showed how "technologically enabled sharing-economy networks are changing the way humans trade and collaborate." Using a new online "sharing" game, they explored quantitative determinants of human sharing strategy, that is, the mathematical nature of the effects of network structure. For the purpose of this chapter, I understand that sharing networks offer the possibility of altruism spreading in a multiplicative manner across a social net.

Thus, there are times when an individual stimulated to act altruistically could, through spillover, see the effects of their good behavior multiplied. I do not want to use the word "contagion" because I was writing this during the corona virus pandemic, but an explosive multiplicative spread of altruistic behavior would seem hugely desirable. And, of course, I have argued that individuals acting altruistically toward other individuals (or groups) can be identified with brain mechanisms that have evolved to favor altruism and empathy. This is a very big deal. It demonstrates that our species has survived, in part, because we are naturally inclined to help our own. Altruism is one of many instinctive social behaviors that are traceable to specific brain mechanisms. This is why brain mechanisms, along with evolution and babies' development, constitute three inextricable strands of how I describe the origins of human social behavior.

Of all the instinctive social behaviors that I have described (e.g., sex, parenting, aggression, and others), altruism is probably the hardest to accept, even though I have mapped onto the brain some precise operations that can cause it. This is because we often tend to think that most people, irrespective of the medical workers devoting themselves to coronavirus, will naturally look for a quid pro quo—some sort of return. That is, we assume that what looks like self-sacrifice could just be a way of seeking promotion, or maybe piling up good-will to be called in like poker chits later in the game.

But this approach misconceives the point of social relations as a zero-sum, transactional series of encounters. The brain does not always work that way. Because it acts automatically, and literally compels us to act—to want sex, to parent, to be altruistic—there may be no time for such "economical" types of calculation. The social instinct celebrates in this book is based on *group* survival, not individual aggrandizement. Thus, when we extend a hand as a greeting or to help, the brain sees another human, not an opportunity to one-up that other human. If we had evolved, always and only, to take selfish advantage of each other, we would likely not be here at all. As it is, we are here because sex, parenting, altruism, and the other social impulses work so well on behalf of us all.

Altruism is thus a legitimate effect of our brain mechanisms, evolved over eons and exhibited even by babies—my three-pronged analysis of prosocial impulses applies here.

In the next chapter, I will discuss social approach/avoidance, the most fundamental dichotomy in all behaviors where humans must react to a stimulus. Neuroscientist Helen Barbas, at Boston University, and I think that social approach versus social avoidance decisions can be understood through a new, straightforward theory of prefrontal cortex outputs through forebrain motor control regions.

New questions

1. Will future research bear out my theory (2015) about how blurring or merging of images in the cerebral cortex contributes to altruistic behavior?
2. What will be the full list of behavioral components of so-called "theory-of-mind" performance?
3. Is there a specific network in our brain that supports "theory of mind" behaviors?
4. Whether or not the answer to question 3 is "yes," do mirror neurons support empathic behaviors, and if so, how?
5. How does our understanding of mechanisms for altruistic behaviors bear on thinking about non-profit organizations?
6. Rensselaer Polytechnic Institute professor Chris Sims (2018) has proposed a new theoretical approach to the universal law of generalization in human perception. Does that bear on our altruistic behaviors toward people who look like ourselves compared to people who don't?
7. To what extent are altruistic, philanthropic behaviors described by philosopher Jeremy Bentham's fundamental axiom to go for "the greatest happiness of the greatest number" of people?

References

Barragan, R.C., et al., 2020. Altruistic food sharing behavior by human infants after a hunger manipulation. Sci. Rep. 10, 1785–1789.
Bartal, I., et al., 2011. Empathy and pro-social behavior in rats. Science 334, 1427–1431.
Bartal, I.B.-A., et al., 2014. Pro-social behavior in rats is modulated by social experience. eNeuro 3, e01385.
Bitsch, F., et al., 2018. The role of the right temporo-parietal junction in social decision-making. Hum. Brain Mapp. https://doi.org/10.1002/hbm.24061.
Carter, R.M., et al., 2012. A distinct role of the temporal-parietal junction in predicting social guided decisions. Science 337, 109–112.
Christov-Moore, L., et al., 2020. Predicting empathy from resting state brain connectivity: a multivariate approach. Front. Integr. Neurosci. 14, 3.
De Waal, F., 1997. Bonobo, the Forgotten Ape. The University of California Press, Berkeley, CA.
De Waal, F., 2009. The Age of Empathy. Three Rivers Press/Random House, New York.
Jennings, J.H., et al., 2019. Interacting neural ensembles in orbitofrontal cortex for social and feeding behavior. Nature 565, 645–651.
Pfaff, D.W., 2015. The Altruistic Brain. Oxford University Press, Oxford.
Rutherford, A., 2019. Humanimal: How *Homo sapiens* Became Nature's Most Paradoxical Creature. The Experiment, London.
Schaafsma, S.M., et al., 2015. Deconstructing and reconstructing theory of mind. Trends Cogn. Sci. 19 (2), 65–72.
Shakya, H.B., et al., 2017. Exploiting social influence to magnify population-level behaviour change in maternal and child health: study protocol for a randomized controlled trial of network targeting algorithms in rural Honduras. BMJ Open 7 (3). https://doi.org/10.1136/bmjopen-2016-012996.
Shirado, H., et al., 2019. Resource sharing in technologically defined social networks. Nat. Commun. 10 (1). https://doi.org/10.1038/s41467-019-08935-2.
Sims, C.R., 2018. Efficient coding explains the universal law of generalization in human perception. Science 360, 652–657.
VanderWeele, T.J., Christakis, N.A., 2019. Network multipliers and public health. Int. J. Epidemiol. https://doi.org/10.1093/ije/dyz010. pii: dyz010.
Wilson, E.O., 2019. Genesis: The Deep Origin of Societies. Liveright, London.
Wrangham, R., 2019. The Goodness Paradox: The Strange Relationship between Virtue and Violence in Human Evolution. Pantheon, New York.
Wu, J., et al., 2018. Exposure, hazard, and survival analysis of diffusion on social networks. Stat. Med. 37 (17), 2561–2585. https://doi.org/10.1002/sim.7658.

Further reading

Christakis, N.A., 2019. The Evolutionary Origins of a Good Society. Little Brown, Boston, MA.
Dawkins, R., 2006. The Selfish Gene. Oxford University Press, Oxford.
Nowak, M., Highfield, R., 2011. Supercooperators: The Mathematics of Evolution, Altruism and Human Behavior. Free Press, New York.
Wilson, E.O., 2008. The Superorganism. WW Norton, New York.

CHAPTER 9

Failure to approach, autism

If altruism represents a high point in human social relations, then the failures of social approach in autism represent something that is not just "different," but worthy medical and scientific attention. With autism, *not* socially engaging is a defining characteristic. Autistic children often do not exhibit an ordinary interest in other people, or may act in ways that keep other people at a distance (Fig. 1).

In his observations about limitations on sociality, relevant to autism, the renowned behavioral biologist Irenaus Eibl-Eibelfeldt (1989) stated: "in higher vertebrates, relationships with conspecifics are marked by a *distinct ambivalence*. The conspecific is, on one hand, a partner that is sought. On the other hand, the tendency to bond is opposed by aggressive and fearful tendencies that encourage maintaining a distance. The conspecific bears signals releasing both friendly approach and fear-motivated withdrawal or aggression" (p. 170). In light of Eibl-Eibelsfeldt's words, can children with the social avoidance of altruism be brought into society as much as possible?

To answer this and related questions, which broach wide-ranging issues of education, rehabilitation, alternative employment, housing and, in some instances, perpetual supervision, one tries to understand the neurobiology of autism.

When autism does occur, it manifests early, i.e., during childhood. Accordingly, I introduced the subject in Chapter 2 regarding babies' development. There I listed nine independent causes for the huge sex difference in autism: Diagnoses in boys about four times as often as in little girls.

Ordinarily, children achieve a balance between approach and withdrawal (avoidance) using understandable emotional rules. But a considerable number of autistic children almost always choose withdrawal. A lot of work has been done on why this is so, an especially difficult task since no two autistic children necessarily exhibit the same range of behaviors. Medical

Origins of Human Socialization
https://doi.org/10.1016/B978-0-323-85861-8.00002-5

Fig. 1 At his birthday party, a child with autistic symptoms might ignore his guests and instead read a book in a corner.

observations of children in the last century led to the concept of Autism *Spectrum* Disorder (ASD) which, in the name "Spectrum," recognizes large variances among children who are socially aloof.

Boston University professor Helen Barbas, chief of the Neural Systems Laboratory there, and I have conceived of mechanisms for the social avoidance in autism as a special case of a dichotomy between two types of fundamental motor response regulations managed by neurons in the prefrontal cortex and the motor control neurons in the basal forebrain: Approach versus Avoid. We wrote about this concept and its supporting scientific literature for Trends in Neuroscience (Pfaff and Barbas, 2019), from which this chapter is adapted. I should note that thinking about autism is important, both emotionally and economically. As a neurodevelopmental disorder with serious lifelong consequences, autism is receiving considerable attention from neuroscientists and geneticists. The lifetime cost for taking care of an autistic child averages in the millions of dollars, not counting lost productivity. The uncertainty and agony for the child's parents can be difficult to witness.

In the theory that I developed with Prof. Barbas, the intention to approach or avoid is seen as a concrete example of "embodied cognition" and explained by dual opposed pathways through the motor control neurons in the basal forebrain. Contrasting these two types of pathways, the "direct" cortical pathway that innervates neurons in the motor control neurons in the basal forebrain that express dopamine D1 receptors, is proposed to facilitate social approach, while the "indirect" pathway, signaling through dopamine D2 receptors, is thought to promote social avoidance. As highlighted by the results of geneticist and psychiatrist professor Matthew State at the University of California, San Francisco Medical School (Willsey et al., 2013), mutations of genes expressed in Layer 5 prefrontal cortical neurons that project to these motor control neurons in the basal forebrain may be key to understanding a bias for avoidance responses in autism.

Then, differences in hormonal sensitivities and the *timing* of genesis of prefrontal cortical neurons between the sexes, among other factors, help explain the predominance of autism in boys. In addition, mutations that affect the *timing* of development, and of mutational insults affecting different *time* points during gestation may help to explain the tremendous inter-individual variability along the autism spectrum.

Our theory of mechanisms plausibly affected during brain development in autism is based on a wealth of knowledge of pathways associated with social behavior, from prefrontal cortex to the motor control neurons in the basal forebrain, and is based on contributions by hundreds of laboratories. As noted, we regard failure of social approach in autism to be a special case of the fundamental dichotomy in behavior: to approach vs. avoid.

Embodied cognition

Accumulating evidence indicates that psychological intentions for action can be represented in activities of neurons in high-level motion-regulatory areas. Bringing the body to mind is the ultimate expression of intentional attitude (Damasio, 2010). The concept of "embodied cognition" attributes the birth of the mind to the brain's control of organized movement (Llinas and Roy, 2009), in effect, translating the abstract into the concrete. There is an analogy in Adams (1974), where Albert Einstein's "mechanisms of thought" are said not to depend on language, but on images "of visual or muscular type."

Embodied cognition has been invoked to understand social approach (Price and Harmon-Jones, 2010), mirror neuron physiology

(Caramazza et al., 2014) and enhanced cognitive training (Ghio et al., 2018, Moeller et al., 2012), because "motor integration affects cognitive integration" (Mullet et al., 2014). The concept can be applied now to further understand how the most fundamental behavioral bifurcation—approach versus avoidance—can be realized in the pathway from prefrontal cortex to medium spiny neurons, influenced by specific gene expression during development.

Dual prefrontal pathways may mediate approach versus avoidance

For conceiving in a new way how we make such fundamental, behavior-regulating decisions as approach vs. avoid, Barbas and I consider the impact of prefrontal cortical pathways on Medium Spiny Neurons of the motor control neurons in the basal forebrain, which have been implicated in decisions on cost-benefit trade-offs (Friedman et al., 2015, 2017). To frame a working hypothesis of the approach/avoidance decision we begin by applying the concept of embodied cognition to the governance of approach vs. avoidance mechanisms through prefrontal cortex to medium spiny neurons pathways. This approach helps clarify how patterns of genetic mutations in prefrontal cortex pyramidal neurons linked to autism spectrum disorder (Willsey et al., 2013) can bias medium spiny neurons' systems away from social approach and towards social avoidance, as seen in autism.

The main hypothesis put forward here is that neurons in prefrontal cortex that are associated with social functions influence medium spiny neurons, which express dopamine D1 receptors in the "*direct path*" to foster social approach. In contrast, medium spiny neurons expressing dopamine D2 receptors through the "indirect path" promote social avoidance. Malfunctioning of this system, especially in males, contributes to the social avoidance in autism.

Thus, we apply data about a fundamental division among behaviors—approach versus avoidance—to social behaviors, because failures of social approach constitute a key diagnostic feature of autism.

Approach versus avoidance within the prefrontal social network

To understand one aspect of the approach/avoidance dichotomy, consider the special relationship of prefrontal cortex with the motor control neurons in the basal forebrain. Whereas the entire cortex projects to basal forebrain motor control pathways, the frontal cortex, which encompasses the prefrontal cortex, and motor/premotor cortices, preferentially receives the bulk of the output of basal forebrain motor control pathways through the thalamus (reviewed in Barbas, 2015). Widespread cortical information is thus available to the frontal cortices, the chief arbiters for decision and action. Among these prefrontal cortex regions that receive privileged information from the output of basal forebrain motor control pathways are those associated with affective behavior and social functions, namely the anterior cingulate cortex, and the orbitofrontal cortex. This circuit implicates the frontal lobe in the process of embodied cognition.

The pathways from the cortex to basal forebrain motor control pathways also show remarkable specialization for eye movements towards stimuli with affective significance (Hikosaka et al., 2019). These specialized pathways involve the superior colliculus to orient the eyes to desired objects in a pathway akin to the direct pathway, or avoid orienting to undesirable objects through the indirect pathway.

Our ideas spring from a rich neurobiological literature. Classical (DeLong, 1990) and modern studies have described two neuronal pathways with opposite effects on movement across mammalian species (Alexander and Crutcher, 1990; Kravitz et al., 2010, 2013; Parker et al., 2018): A "direct" pathway from the cortex to basal (basal forebrain motor control pathways) facilitates movement. An "indirect" pathway from cortex to basal forebrain motor control pathways opposes movement. Motor control neurons in the basal forebrain disorders reveal specific deficits after disruption in these dual opposing pathways (Albin et al., 1989).

The cortical pathways to the input nuclei of basal forebrain motor control pathways, called the caudate and putamen, are topographic. Anterior (frontal) cortices project to the caudate head, and progressively more posterior cortices project to the body of the caudate, putamen and the tail of the caudate nucleus. The distinction of the direct and indirect pathways is first determined by innervation of the caudate or putamen medium spiny neurons s that express D1 receptors in the direct pathway and medium spiny neurons s that express D2 receptors in the indirect pathway.

Now we describe the forebrain systems for social approach and avoidance in a manner which avoids the baffling series of names for the subdivisions of non-cortex motor control pathways.

Direct pathway for approach

Another distinction between pathways which might foster social approach versus those for avoidance is based on the downstream projection to the output of basal forebrain motor control pathways and then to the thalamus.

The "direct" pathway projects to the internal segment of basal forebrain motor control pathways, which in turn disinhibit the thalamus and, by virtue of thalamic neuronal projections widely across the cortex, facilitate initiation of voluntary movement. In contrast, the medium spiny neurons of the "indirect" pathway innervate a different part of the basal forebrain motor control pathways, which project to the excitatory sub-thalamic nucleus, which innervates the thalamus. The "indirect" circuit also has additional components which innervate the thalamus. All of these have the same effect: to prevent movement (Fig. 2).

A similar situation applies to our looking at another person and treating them either as someone to approach or to avoid. In this special case of eye movements, chief of the Section on Neural Networks at the National Institutes of Health, Okihide Hikosaka, and his colleagues propose that pathways from part of our basal forebrain motor control pathways are associated with long-term memory of the value of stimuli, mediated through direct and indirect pathways. The wiring of the dual pathways is similar to the classical pathways through basal forebrain motor control pathways, but involve the midbrain's superior colliculus which controls movement of the eyes. The direct pathway features a net excitation, achieved by a simple disinhibition, allowing the eyes to move towards good objects, objects that deserve approach. The "indirect" pathway has an extra projection through forebrain motor control pathways, an extra synapse which achieves the opposite effect, *avoidance* of directing the eyes to an object known to be "*bad*."

There is an important direct pathway from the motor-controlling region which projects to the midbrain in a pathway which has a role in voluntary eye movements in tasks associated with short term memory. This pathway appears to be special for choosing visual objects. Autism-related perturbations through the basal forebrain motor control pathways could account for low or otherwise dysregulated eye contact.

Indirect pathway for avoidance

Based on the canonical basal forebrain motor control pathways, our interpretation is that the most anterior motor control source in the forebrain has still another *indirect* pathway, which courses through downstream synapses to enter the midbrain and which would inhibit the midbrain's superior colliculus and prevent eye movements. It is also conceivable that pathways associated with attention in the anterior cingulate cortex (reviewed in Barbas et al., 2013) can work through the direct pathway to focus attention on

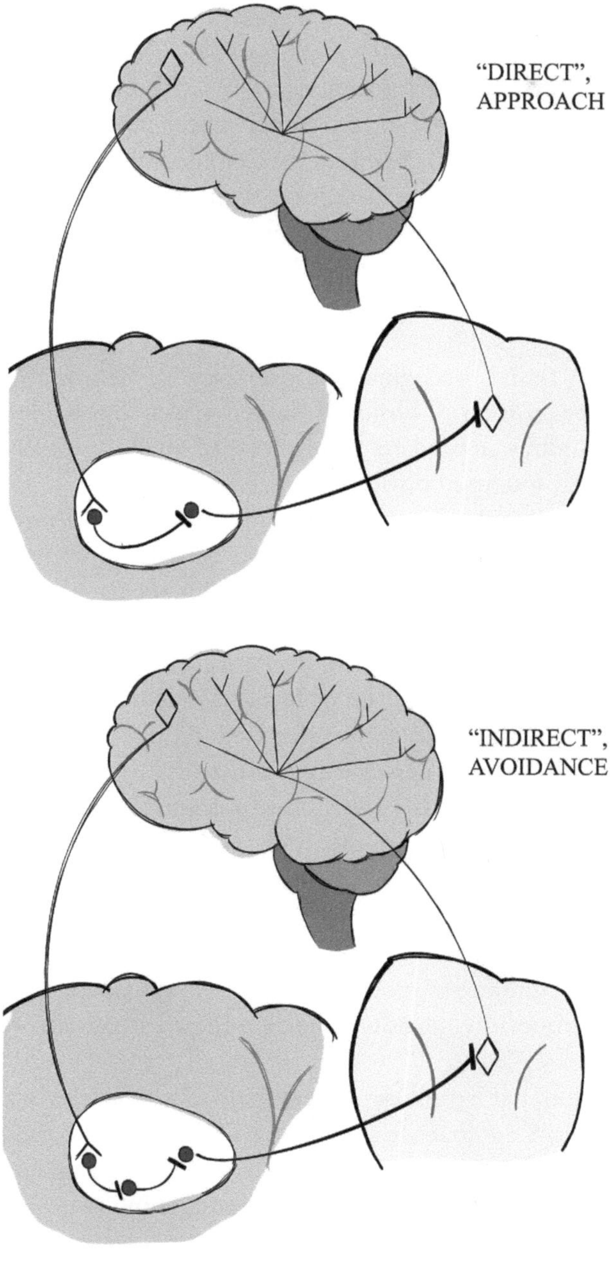

Fig. 2 Regulating the balance between social approach and social avoidance. Frontal cortex outputs routed through basal forebrain motor control regions can be flipped from approach to avoidance by the addition of an inhibitory synapse (−**1**). Explained in Barbas and Pfaff (2019).

rewarding stimuli, or to avoid non-rewarded stimuli or eye contact and social interactions.

The universal cortical projection to basal forebrain motor control pathways thus includes the orbitofrontal cortex and anterior cingulate cortex (Haber et al., 2000; Selemon and Goldman-Rakic, 1985; Yeterian and Pandya, 1991), which are associated with social processes (Barbas and García-Cabezas, 2017). These projections to medium spiny neurons s thus provide the circuit basis for approach/avoidance decisions in a social context, including directing attention to social events. We note that long ago, in 1959, MIT professor Jerry Lettvin and colleagues ("What the Frog's Eye Tells the Frog's Brain") recognized that neurons in the frog midbrain were dedicated to acquisition of positively valued visual targets, like flies to eat. Their early finding attested to the behavioral significance and enduring relevance of this system in phylogeny.

The hypothesis advanced here is supported by work with primates, whereby there is a direct pathway through basal forebrain motor control pathways and the midbrain's superior colliculus mediates eye movements towards objects that are good, while an indirect pathway mediates avoidance of gaze to objects known to be bad. In the respective pathways, eye movements are perturbed after injection of antagonists to dopamine D1 receptors or D2 receptors at the input sector of basal forebrain motor control pathways. This evidence strengthens the association of the indirect pathway with "bad" objects (Kim et al., 2017), including presumably social interactions to be avoided. These findings with primates can be related to the importance of the indirect pathway for social avoidance, a behavior that is characteristic of many individuals with autism.

As a side point, we note that the dual pathways both increase their activity at the very beginning of movements, suggesting that collaborative action is needed for smooth regulation of flexible behavior patterns (reviewed in Bostan and Strick, 2018).

Thus, the simplest hypothesis offered here states that the direct pathway serves to facilitate approach responses, and the indirect pathway opposes them. This formulation explains approach/avoidance regulation based on known neuroanatomy and neurochemistry, suggesting that the dual circuit synergism is disrupted when one of the pathways is preferentially affected.

Social approach versus social avoidance

As we have suggested, the brain's regulation of social approach constitutes a special case of the approach/avoidance dichotomy. Prefrontal cortical neurons are intimately involved in the control of social behaviors: (i) at the level of circuits; and (ii) in expression of genes and markers associated with social interactions. The extension of the approach avoidance dichotomy to social behavior is a parsimonious hypothesis that can be tested in future studies. This approach leads us to detailed consideration of such circuitry and gene expression:

(i) At the circuit level, neurons in medial prefrontal cortex in rodents, and in medial (anterior cingulate cortex) and orbitofrontal cortex areas in the more specialized cortex of primates (Zikopoulos et al., 2017), have strong connections with the amygdala, the hypothalamus and other subcortical structures associated with social behavior (reviewed in Barbas, 2000). In macaque monkeys, some neurons in medial prefrontal cortex are activated exclusively by watching social interactions (Sliwa and Freiwald, 2017), the impending reward of a social partner, or social investigation. Recent evidence in voles has shown low-frequency electrophysiological coherence between medial prefrontal cortex and the reward circuitry. Collectively, neurons engaged while animals witness social interactions could provide inputs to medium spiny neurons to favor social approach or avoidance.

(ii) At the level of expression of specific markers, we note that neurons of the rat prefrontal cortex express oxytocin receptors (Mitre et al., 2016; Oettl et al., 2016), activated by the neuropeptide oxytocin, known for supporting prosocial behavior (Garrison et al., 2012; Jurek and Neumann, 2018) prefrontal cortex (Nakajima et al., 2014). In the rodent mammalian system oxytocin signaling permits a range of social behaviors by buffering regulatory pathways from the disruptive effects of stress (McCarthy et al., 1991), an hypothesis that has received physiological and clinical support. In rats and mice oxytocin plays an important role in supporting social behaviors crucial for reproduction (reviewed in Jurek and Neumann, 2018; Pfaff, 2017), and fosters social recognition (Choleris et al., 2003, 2006). Silencing neurons that express the oxytocin receptor or interfering with the oxytocin receptor in medial prefrontal cortex abolishes the social interest that female mice exhibit during the sexually receptive phase of the estrous cycle (Nakajima et al., 2014). Further, findings of a role of estrogen hormones in physiological signaling and robust strengthening of oxytocin receptors and their transcription are consistent with the fact that females do not succumb to the social difficulties of autism as frequently (about 20% of diagnoses) as males (about 80% of diagnoses).

> By contrast, in male mice, corticotropin-releasing-hormone-binding protein (CRHBP), an antagonist of the stress hormone corticotropin releasing hormone (CRH), is expressed in oxytocin receptor expressing neurons in prefrontal cortex. Production of CRHBP blocks the CRH-induced activation of a specific subclass of prefrontal cortex interneurons and produces an anxiolytic effect (Li et al., 2016).

The possibility that abnormalities in oxytocin systems contribute to the development of social symptoms in autism has been considered (Gregory et al., 2009; Johnson and Young, 2017; Torres et al., 2018), along with evidence of abnormalities in the frequency of DNA methylation in two specific regions of the oxytocin receptor promoter. However, the presence of Oxytocin receptor neurons in primates has not been demonstrated at the cortical level as it has been in rodents, although oxytocin binds with high affinity to vasopressin receptor V1ARs, which may mediate oxytocin action in a variety of mammals, including primates (Everett et al., 2018; Freeman et al., 2014; Loup et al., 1991).

Expanded regions analogous to the rodent medial prefrontal cortex, namely the anterior cingulate cortex, and the orbitofrontal cortex in primates are engaged in social functions, and maintain robust connections with cortical and subcortical limbic structures, such as the hypothalamus and the amygdala. In humans, the pathways below the anterior cingulate cortex and orbitofrontal cortex are altered in distinct ways in autism (Zikopoulos and Barbas, 2010). The main hypothesis here predicts that neurons in these prefrontal cortex regions, in particular, project to, and facilitate, activity in the direct pathway, fostering social approach. By contrast, prefrontal cortex pathways acting via D2 dopamine receptors in neurons in the deep layers of anterior cingulate cortex and/or orbitofrontal cortex project to neurons with D2 receptors in basal forebrain motor control pathways that foster social avoidance.

Prefrontal cortex-striatal pathways in autism

The ideas presented here provide one of many possible routes to help explain the problems of social communication seen in children with autism, namely the causal route of a prefrontal cortex problem regarding signaling through medium spiny neurons. This explains, in part, a major diagnostic feature of autism, i.e., a deficit in social relations, which to a substantial degree is characterized by a failure of social approach.

Accumulating evidence through the present has identified hundreds of genes that could be implicated in the etiology of autism (Coe et al.,

2019; Parikshak et al., 2016; Ramaswami and Geschwind, 2018; Sestan and State, 2018; Turner and Eichler, 2019). Epigenetic mechanisms such as DNA methylation, histone acetylation and histone methylation have also been drawn into causal arguments. In the context of prefrontal pathways to medium spiny neurons Barbas and I asked: what are the consequences of altered gene expression for the forebrain circuits? The simplest hypothesis states that some individuals with autism suffer from dysregulation of prefrontal signaling through medium spiny neurons (Peca et al., 2011).

The social importance of the link from prefrontal cortex projection neurons to basal forebrain motor control pathways is strengthened by genetic evidence (Willsey et al., 2013), indicating that mutations associated with prefrontal layer 5/6 pyramidal projection neurons in development of the human brain are implicated in the etiology of autism. Layer 5 neurons project to basal forebrain motor control pathways and the amygdala and, along with layer 6 neurons, to the thalamus. Beginning with high-confidence autism-related "seed genes" and measuring gene expression in several human brain regions across development, comprehensive coexpression networks led to the discovery of "a key point" of genetic expression in midfetal layer 5/6 prefrontal (and other) cortical neurons. In view of the fact that a common diagnostic feature of autism is social avoidance, it is logical to propose that the mutations reported (and perhaps others as well) alter the balance of activity in prefrontal cortex circuits through projections to medium spiny neurons s in such a manner as to discourage social approach and provoke social avoidance.

How would preferential insult to one pathway affect social interactions? Mouse models of autism have specifically shown abnormal medium spiny neurons activity in the indirect pathway (Peca et al., 2011; Wang et al., 2017). Following up from studies about deficits in prediction in autism, new evidence further implicates the motor control neurons in the basal forebrain in animal models of autism (reviewed in Subramanian et al., 2017). Portmann et al. (2014) explored the 16p11.2 deletion widely associated with autism in patients, by analyzing the corresponding region on mouse chromosome 7F3. They found an increase in the number of medium spiny neurons s expressing the dopamine D2 receptor gene. Interestingly, in primates dopamine receptors show preferential laminar distribution in prefrontal cortex, with D1 receptors expressed in layers 2–3, and D2 receptors in layers 5–6. In prefrontal cortex dopamine acts through D2 receptors on projection neurons that are directed to medium spiny neurons s that express D2 receptors. Thus, both physiological and genetic studies lead to the concept that abnormalities in the indirect pathway, involving medium spiny neurons that express the dopamine D2 receptor gene, could help explain enhanced social avoidance.

Genetics in autism

Over the years, heritability estimates of Autism Spectrum Disorders (ASD) have ranged as high as 90%, and the evidence has been comprehensively reviewed. For example, organized concepts of genetic "models" of autism have been proposed by Berg and Geschwind (2012). And suffice it to say that hundreds of genes can be involved, with mutations either in their coding regions or their non-coding regulatory elements (reviewed by Turner and Eichler, 2019).

Epigenetics

We suspect that changes in DNA sequence in coding regions comprise only the first-examined and perhaps the simplest neuronal nuclear biochemical events crucial to the biological etiology of autism. In other fields of cell biology, regulatory *epigenetic* post-translational modifications—"epigenetic," meaning above and beyond DNA nucleotide bases changes, themselves—such as histone acetylation, histone methylation and DNA methylation are analyzed in great detail (reviewed, Allis and Jenuwein, 2016; Sabari et al., 2017). One can envision hierarchies of molecular controls over synthetic events in the nerve cell nucleus, hierarchies whose nature, logic and timing are not yet established.

Most important, epigenetic changes related to autism envision a clear path for prenatal and neonatal experience to render changes in social behavior-related gene expression in the brain. Some of those changes could be causal to autism. Here are just two of many possible examples.

Chromatin regulators. Some of the high confidence autism mutations converge around the concepts of synaptic function but also center on the idea of chromatin regulators. Chromatin remodeling involves histone-modifying enzymes ("writers") and chromatin remodelers that recognize specific histone post-translational modifications and orchestrate their effects on chromatin. That is, chromatin regulators typically include "writer proteins" that establish and maintain epigenetic marks, "eraser proteins" that remove modifications, and "reader proteins" that bind and interpret epigenetic information.

DNA methylation. DNA methylation is one of the major epigenetic regulators of gene expression and associated phenotypes. Methylation patterns in genomic DNA are generated during embryogenesis and early fetal development and are altered throughout life in response to endogenous or exogenous environmental signals. Recently, differentially methylated variants (DMVs) at specific CpG sites or differentially methylated regions (DMRs) have been investigated in ASD individuals (Antequera and Bird, 2018), using various types of tissues, including lymphoblastoid cell lines, whole blood, and brain tissue samples (Ladd-Acosta et al., 2014; Nardone et al., 2014). It has been postulated that differentially methylated genes are associated with several biological functions, including gene transcription, nervous system development, and other biological mechanisms implicated in ASDs.

Sex differences in autism

A leading autism researcher, Daniel Geschwind, MD, a professor at the University of California Los Angeles, wrote that "autism spectrum disorders have long been noted to affect many more males than females and, accordingly, sex-specific factors have been hypothesized to increase male risk…" (Werling et al., 2018). Likewise, the British psychologist Simon Baron-Cohen blamed autism on "the extreme male brain" (2002). But what is the actual reason (as opposed to simply citing male identity) for the predominance of autism in males? What are the genetic contributors affecting the brains of autistic individuals?

In the midst of the heterogeneity of symptoms among autistic children, can we account for the most stable feature of autism, the large difference in occurrence between the sexes? As I introduced in Chapter 2, the most obvious causal factor for sexual differentiation of the brain is the exposure of the male brain to high levels of testosterone (Fig. 3) in early life (Pfaff et al., 2011), but at least three other genomic factors may play a role: the complex details of X-inactivation, the parent-of-origin allelic imprinting, and the inactivated X as a heterochromatin sink (Schaafsma and Pfaff, 2014). Even the initiation of genetic recombination shows female/male differences. A "3-hit" theory of the sex difference in autism—(i) genetic predisposition, (ii) maternal stress, and (iii) maleness—made some correct predictions in a mouse experiment based on the discovery of CNTNAP2 as an autism susceptibility gene (Schaafsma et al., 2017a).

Sex differences in autism likely depend on circuits chemically-coded by specific hormone-dependent neuromodulators, as illustrated by

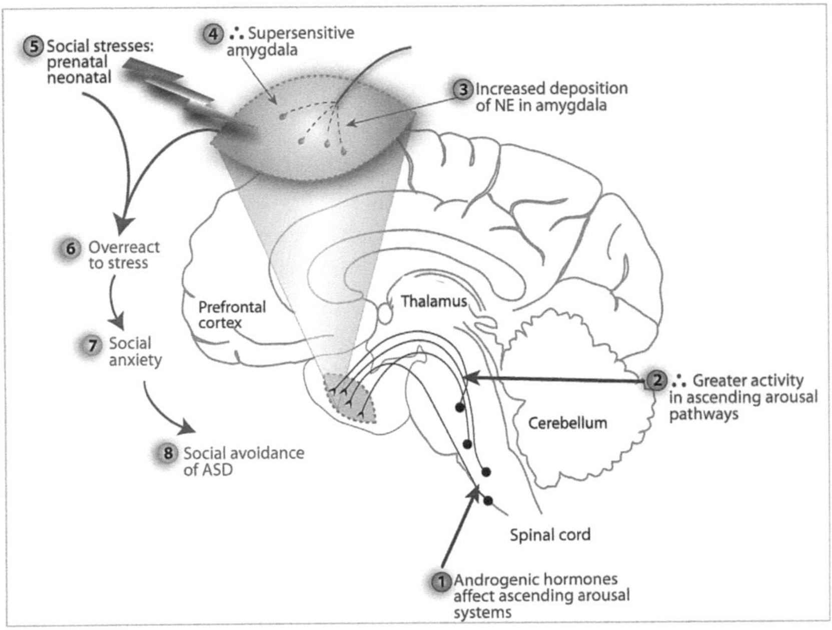

Fig. 3 Illustration of the "3-hit" theory of the sex difference in autism: (i) sex hormones, (ii) stress and (iii) genetics. Due to the influence of androgenic hormones (like testosterone) on brainstem arousal pathways, a super-sensitive amygdala over-responds to stress, including social stress. These hormonal and environmental influences are compounded by genetic mutations associated with autism.

estrogen-dependent neurons (Pfaff and Keiner, 1973; Cottingham and Pfaff, 1986) that likely alter connectivity and synaptic function (Bargmann, 2012). For example, hormonal regulation can be different between males and females.

To relate the sex differences in autism to our forebrain circuit hypothesis (Chapter 10), the primary link likely is in prefrontal cortex. This hypothesis is supported by sex differences in prefrontal cortex, at least in rodents (reviewed in Marrocco and McEwen, 2016), and expression of both estrogen and androgen receptors (Westberry and Wilson, 2012; Low et al., 2017), all of which could participate in developmental programming. Molecular pharmacologic investigations have drawn medial prefrontal cortex into the regulation of social behaviors of female rats, including a "functional reorganization" during the estrous cycle, emphasizing the binding of the immediate early gene Egr1 to synapse-related genes, which was previously related to social anxiety. Overexpression of extracellular signal regulated kinase 2 in medial prefrontal cortex through microinjection of a modified herpes simplex virus increased the duration of social interactions by almost 50%. In medial prefrontal cortex, the effectiveness of specific types of GABAergic interneurons is thought to differ between female and male mice. Thus, the molecular and physiological consequences of sex differences in medial prefrontal cortex in rodents contribute to our understanding of the etiology of autism.

Is there a female protective effect? As noted by Daniel Geschwind, another possibility is that females are protected from the factors that cause autism, whether those factors are genetic or environmental. Accordingly, Werling et al. (2018) state: "the premise of a female protective effect model is simple: if a female-specific factor protects females across the board from reaching the threshold for autism diagnosis then those female who are affected are likely carrying a greater etiologic load than affected males who lack this female-specific protective factor." Geschwind quotes data from Robinson et al. (2013), who reported that they "compared sibling autistic traits between female and male probands, who were identified as children scoring in the top 90th and 95th percentiles of the population autistic trait distributions. In siblings of female probands above the 90th percentile had significantly more autistic impairments than the siblings of male probands above the 90th percentile. The siblings of female probands above the 90th percentile also had greater categorical recurrence risk. Results were similar in probands above the 95th percentile. This finding, replicated across two nationally-representative samples, suggests that female sex protects girls from autistic impairments and that girls may require greater familial etiologic load to manifest the phenotype."

I am not entirely convinced by this indirect reasoning, and see two problems in the conclusions regarding a potential female protective effect (FPE). First, logically, can we truly divorce a female protective effect from a male-specific liability? Second, to date, there is no firm genetic locus reported for a female protective effect. Nevertheless, let's look at the data.

In the 1980s Tsai et al. found that autistic girls have more relatives with autism or certain language impairments, on average, than do boys with the condition (Tsai et al., 1981, 1983). This finding hinted that girls need to inherit more factors related to autism than boys in order to be on the autism spectrum. Several studies have supported Tsai's observation, by raising the possibility that the sex bias that is consistently observed in ASD may, in part, be the consequence of a female protective effect (FPE), thus reducing the incidence in females (Jacquemont et al., 2014; Sanders et al., 2011; Dong et al., 2014; Fombonne, 2009; Werling and Geschwind, 2015; Levy et al.,

2011; Messinger et al., 2015; Robinson et al., 2013; Gockley et al., 2015), but one study could not find evidence of a single FPE locus (Gockley et al., 2015).

One single-gene disorder that has an intriguing and complex relationship with autism is the Fragile X syndrome (FXS). The disorder is caused by the silencing of a single gene on the long arm of the X chromosome at Xq27.3, and reductions of social interactions are included among its symptoms. The gene affected is the RNA-binding protein Fragile X Mental Retardation Gene-1 (FMR1) where there is a triplet repeat expansion in the 59 UTR of the FMR1 gene to over 200 repeats. The pre-mutation frequency is 11.7 (95% CI: 6.0–18.7) per 10,000 males and 34.4 (95% CI: 6.3–83.3) per 10,000 for females (1:855 and 1:291, respectively) in the total population. Yet, incidence of the disorder itself is about 1.4 per 10,000 males and 0.9 per 10,000 females have FXS (Loesch et al., 2007; Philofsky et al., 2004; Zang et al., 2009; Vasilyev et al., 2015; Hagerman, 2008; Hunter et al., 2014).

Therefore, these comparisons provide a good example of a FPE where, even though mutation frequency is much higher in females, they present with a lower incidence of the actual disorder and milder symptoms than males (Keysor and Mazzocco, 2002).

This Fragile X permutation (a mutation which increases the risk of having an affected child) results in loss of function through transcriptional gene silencing and subsequent loss of the fragile mental retardation protein (FMRP). This single gene disorder offers an interesting model by which to explore the functions of FMRP regulation particularly in KH2-mediated RNA binding and in synaptic plasticity, with the potential for using it as a therapeutic target for patients with FXS (Darnell and Klann, 2013; Darnell et al., 2001, 2009, 2011; Grigsby, 2016).

The broadest perspective on genetic determinants of autism would note that George Beadle and Edward Tatum (1941) working with the slime mold Neurospora won the Nobel prize for their "one gene/one enzyme" concept, each gene coding for an enzyme, explaining the slime mold's cell biology. In the field of autism research, we are likely dealing with *patterns* of genes linked to *patterns* of human behavior.

Development and the potential for heterogeneity in autism

In autism, what accounts for the tremendous heterogeneity of symptoms from patient to patient? One possible source is that the genetic mutations which predispose to autism are manifest at different developmental stages of the brain. As a result, the endophenotypes in autism due to those mutations would be different. This would result in heterogeneity.

Since autism is a developmental disorder, this raises the question of the mechanisms that may help explain the preferential involvement of some areas over others in autism, the significant sex differences, and inter-individual variability. In this context, the limbic anterior cingulate cortex and orbitofrontal cortex regions, which have key roles in social interactions, develop early in ontogeny (reviewed in Zikopoulos and Barbas, 2010). Pathways from the deep layers—which develop first—project to medium spiny neurons (reviewed in Graybiel, 1991, 1995). White matter tracts located below frontal cortical regions are disrupted in adults with autism (Zikopoulos and Barbas, 2010; Zikopoulos et al., 2018a), and show persistent and increasing aberrant development from childhood to adulthood (Zikopoulos et al., 2018b).

It is not surprising that preferential involvement of some brain areas in neurologic and psychiatric diseases may be traced to their damage during development. This hypothesis helped unify two distinct changes in axons in the brains of adults with autism in two prefrontal cortex regions. Specifically, below the orbitofrontal cortex the myelin was thinner in autism (Zikopoulos and Barbas, 2010). On the other hand, below anterior cingulate cortex there was a significant increase in thin and medium diameter axons, which link closely situated areas within the frontal cortex, as axons branch when they approach their targets.

Different timing, different symptoms. Are autism-relevant axonal changes in these regions independent or related? To help explain the distinct differences in axons below orbitofrontal cortex and anterior cingulate cortex it is necessary to consider, as an example, the interplay of the developmental processes of neuronal migration, axon growth and branching, and myelination, which proceed in that order and partially overlap in time (Sestan and State, 2018). Thus, expression of an axon Growth-Associated Protein (GAP-43) and myelin-related proteins are mutually antagonistic: when axons grow rapidly myelin proteins are inhibited; when the expression of GAP-43 wanes, myelination begins. Clearly, the exact *timing* of the impacts of any given autism-related mutation will dictate the behavioral effects of that mutation.

Elaborating on this timing issue, after development, GAP-43 is no longer expressed in the cortex in noticeable amounts, except for a few limbic areas, including the anterior cingulate cortex. In anterior cingulate cortex neuronal migration is early in rhesus macaques, as is neuronal maturation in human fetal tissue (Barbas and García-Cabezas, 2016). Myelination is very late in anterior cingulate cortex, consistent with the high expression of GAP-43 in typical humans. But in the brains of adults with autism, significantly more axons below anterior cingulate cortex express GAP-43 (Zikopoulos and Barbas, 2010). This feature helps explain higher branching of axons, and hence the presence of many medium and thin axons below anterior cingulate cortex in adults with autism.

In comparison with anterior cingulate cortex, orbitofrontal cortex neuronal migration proceeds a bit later, GAP-43 levels are lower, and myelination begins earlier. The interplay of these developmental events appears to weaken myelination but leaves axon branching unaffected, as summarized in a developmental model based on available evidence in primates.

In a similar vein, as noted in the Barbas/Pfaff paper (2019), mutations in diverse genes associated with autism *may perturb development at different time points* and produce variable effects on neuronal migration, extension of axons that connect areas, synaptogenesis and myelination. A developmental model for autism thus may be powerful for helping understand the heterogeneity of autism. Many autism related genes have in common a disturbance of developmental events (e.g., Coe et al., 2019; Sestan and State, 2018; Zikopoulos and Barbas, 2010). Based on the type of gene, or combination of genes involved, it is highly likely that disturbances in development are also variable in time, affecting clusters of neurons as they migrate and extend their axons, with the likely prospect that connections also vary across the autism spectrum.

Similarly, sex differences in the timing of development of areas may provide an additional mechanism to help explain the higher incidence of autism in males. A future challenge will be to compare the timing of development of prefrontal cortex in males and females and relate the sex differences in autism to key genetic evidence (Brandler et al., 2018; Willsey et al., 2013), namely mutations in development of prefrontal layer 5/6 projection neurons. The derivation of medium spiny neurons from stem cells is a complex process, involving over 30 neuroanatomical patterning factors, and would seem to offer multiple chances of mistakes that lead to medium spiny neurons dysregulation and behavioral deficits (Tsunemoto et al., 2015).

Every early genetic study provided evidence for one-gene/one-enzyme concept in simple organisms (Beadle and Tatum, 1941). As mentioned above, recent evidence revealed that many gene mutations are associated with autism, providing the basis for multiple gene participation and interactions in

ASD. Phenotypic heterogeneity likely depends on multiple gene mutations that affect prefrontal cortex development at different stages in ontogeny. Again, put more simply, instead of the one-gene/one-enzyme paradigm, we now look for complex *patterns* of gene expression causally linked to complex *patterns* of behavior.

In the next chapter, I will discuss the effect of loneliness and the loss of human companionship. Of course, this will not involve the type of autistic isolation that I have just discussed but, rather, the effect of certain environmental circumstances leading to social deprivation. In discussing the coronavirus, I have already touched on such deprivation. If humans are designed to function in a group, what happens when whole groups devolve into individuals who are unhappy living apart from each other? Even worse, what happens when one or a few members of a group are set apart while the rest go about their business communally? Then next chapter on loneliness will touch on these questions.

New questions

Even as the overriding purpose of this book brings us to look at the causal roots of social approaches in the most comprehensive manner, Helen Barbas and I (2019) explored mechanisms for the opposite behavior pattern: social avoidance.

1. The main hypothesis put forward is that neurons in prefrontal cortex associated with social functions influence medium spiny neurons that express D1 receptors in the "direct path" to foster social approach, as opposed to D2-expressing medium spiny neurons s associated with avoidance. Can current genomic and epigenomic technologies provide insight as to exactly how alterations in the developmental programs in prefrontal cortex neurons lead to an increased risk for autism?
2. Extension of our approach/avoidance idea from non-social to social behaviors is a parsimonious hypothesis for future test. Experimental probing of approach or avoidance behaviors in both social and non-social contexts should provide information as to whether the two types of behavior move in parallel or not.
3. Prefrontal cortex neurons do not function in isolation. Exactly how would functional changes in prefrontal cortex neurons contribute to circuit abnormalities? Initial studies using mouse models of autism have shown abnormal medium spiny neurons activity in the "indirect" pathway. Can further studies uncover "autism-related" changes in the "direct" pathway?

4. The direct and indirect pathways increase their activity at the beginning of movements, suggesting that collaborative action is needed for smooth and flexible behavior regulation. Is this collaborative action in the two paths graded or dichotomous for movement as compared to cognitive and social functions?
5. The most recently developed technical approach to autism involves isolating and sequencing RNA from individual cells in the human cerebral cortex. A team from Arnold Kriegstein's lab at the University of California did this and got results which I would put into two bins. On the one hand, mRNAs involved with the regulation of synaptic signaling piled on data in an already developed field of thought. On the other hand, results indicated autism-related changes in gene expression in cortical microglia, and unexpected finding which needs to be followed up.
6. Estrogen-sensitivity and testosterone-sensitivity provides one, hormonal path to understanding the large sex difference in autism. What about the other causal routes to the sex difference in autism? And do differences in the timing of development of cortical areas associated with social behavior differ between the sexes?
7. Avoidance of social engagement raises the question of why the individual avoids social reward. Recent work from leading neurobiologists at Stanford University (Hung et al., 2017) analyzes mechanisms of social reward and highlights the role of oxytocin in the ventral tegmental area, deep in the forebrain. Dopaminergic neurons there project to a key node of reward circuitry, the nucleus accumbens.

References

Adams, J.L., 1974. Conceptual Blockbusting: a Guide to Better Ideas. Freeman; Distributed by Scribner, San Francisco; New York. xii, 137 p.
Albin, R.L., Young, A.B., Penney, J.B., 1989. The functional anatomy of basal ganglia disorders. Trends Neurosci. 12, 366–375.
Alexander, G.E., Crutcher, M.D., 1990. Functional architecture of basal ganglia circuits: neural substrates of parallel processing. Trends Neurosci. 13, 266–271.
Allis, C.D., Jenuwein, T., 2016. The molecular hallmarks of epigenetic control. Nat. Rev. Genet. 17 (8), 487–500. https://doi.org/10.1038/nrg.2016.59.
Antequera, F., Bird, A., 2018. CpG islands: a historical perspective. Methods Mol. Biol. 1766, 3–13. https://doi.org/10.1007/978-1-4939-7768-0_1.
Barbas, H., 2000. Complementary role of prefrontal cortical regions in cognition, memory and emotion in primates. Adv. Neurol. 84, 87–110.
Barbas, H., 2015. General cortical and special prefrontal connections: principles from structure to function. Annu. Rev. Neurosci. 38, 269–289.

Barbas, H., Bunce, J.G., Medalla, M., 2013. Prefrontal pathways that control attention. In: Stuss, D.T., Knight, R. (Eds.), Principles of Frontal Lobe Functions. Oxford University Press, New York, NY, pp. 31–48.

Barbas, H., García-Cabezas, M.A., 2016. How the prefrontal executive got its stripes. Curr. Opin. Neurobiol. 40, 125–134.

Barbas, H., García-Cabezas, M.A., 2017. Prefrontal cortex integration of emotions and cognition. In: Watanabe, M. (Ed.), Prefrontal Cortex as an Executive, Emotional and Social Brain. Springer Japan, pp. 51–76.

Bargmann, C.I., 2012. Beyond the connectome: how neuromodulators shape neural circuits. BioEssays 34, 458–465.

Beadle, G.W., Tatum, E.L., 1941. Genetic control of biochemical reactions in neurospora. Proc. Natl. Acad. Sci. U. S. A. 27, 499–506.

Berg, J.M., Geschwind, D.H., 2012. Autism genetics: searching for specificity and convergence. Genome Biol. 13 (7), 247. https://doi.org/10.1186/gb4034.

Bostan, A.C., Strick, P.L., 2018. The basal ganglia and the cerebellum: nodes in an integrated network. Nat. Rev. Neurosci. 19, 338–350.

Brandler, W.M., Antaki, D., Gujral, M., Kleiber, M.L., Whitney, J., et al., 2018. Paternally inherited cis-regulatory structural variants are associated with autism. Science 360, 327–331.

Caramazza, A., Anzellotti, S., Strnad, L., Lingnau, A., 2014. Embodied cognition and mirror neurons: a critical assessment. Annu. Rev. Neurosci. 37, 1–15.

Choleris, E., Gustafsson, J.A., Korach, K.S., Muglia, L.J., Pfaff, D.W., Ogawa, S., 2003. An estrogen-dependent four-gene micronet regulating social recognition: a study with oxytocin and estrogen receptor-alpha and -beta knockout mice. Proc. Natl. Acad. Sci. U. S. A. 100, 6192–6197.

Choleris, E., Ogawa, S., Kavaliers, M., Gustafsson, J.A., Korach, K.S., et al., 2006. Involvement of estrogen receptor alpha, beta and oxytocin in social discrimination: a detailed behavioral analysis with knockout female mice. Genes Brain Behav. 5, 528–539.

Coe, B.P., Stessman, H.A.F., Sulovari, A., Geisheker, M.R., Bakken, T.E., et al., 2019. Neurodevelopmental disease genes implicated by de novo mutation and copy number variation morbidity. Nat. Genet. 51, 106–116.

Cottingham, S.L., Pfaff, D.W., 1986. Interconnectedness of steroid hormonebinding neurons: Existence and implications. In: Ganten, D., Pfaff, D.W. (Eds.), Current Topics in Neuroendocrinology. Springer-Verlag, Berlin; New York, pp. 223–249.

Damasio, A.R., 2010. Self Comes to Mind: Constructing the Conscious Brain. Pantheon Books, New York, p. 367. xi.

Darnell, J.C., et al., 2011. FMRP stalls ribosomal translocation on mRNAs linked to synaptic function and autism. Cell 146 (2), 247–261.

Darnell, J.C., Fraser, C.E., Mostovetsky, O., Darnell, R.B., 2009. Discrimination of common and unique RNA-binding activities among fragile X mental retardation protein paralogs. Hum. Mol. Genet. 18 (17), 3164–3177.

Darnell, J.C., Jensen, K.B., Jin, P., Brown, V., Warren, S.T., Darnell, R.B., 2001. Fragile X mental retardation protein targets G quartet mRNAs important for neuronal function. Cell 107 (4), 489–499.

Darnell, J.C., Klann, E., 2013. The translation of translational control by FMRP: therapeutic targets for FXS. Nat. Neurosci. 16 (11), 1530–1536.

DeLong, M.R., 1990. Primate models of movement disorders of basal ganglia origin. Trends Neurosci. 13, 281–285.

Eibl-Eibelfeldt, I., 1989. Human Ethology. de Gruyter, Berlin.

Everett, N.A., McGregor, I.S., Baracz, S.J., Cornish, J.L., 2018. The role of the vasopressin V1A receptor in oxytocin modulation of methamphetamine primed reinstatement. Neuropharmacology 133, 1–11.

Freeman, S.M., Inoue, K., Smith, A.L., Goodman, M.M., Young, L.J., 2014. The neuroanatomical distribution of oxytocin receptor binding and mRNA in the male rhesus macaque (Macaca mulatta). Psychoneuroendocrinology 45, 128–141.
Friedman, A., Homma, D., Bloem, B., Gibb, L.G., Amemori, K.I., et al., 2017. Chronic stress alters striosome-circuit dynamics, leading to aberrant decision-making. Cell 171, 1191–1205-e28.
Friedman, A., Homma, D., Gibb, L.G., Amemori, K., Rubin, S.J., et al., 2015. A corticostriatal path targeting striosomes controls decision-making under conflict. Cell 161, 1320–1333.
Garrison, J.L., Macosko, E.Z., Bernstein, S., Pokala, N., Albrecht, D.R., Bargmann, C.I., 2012. Oxytocin/vasopressin-related peptides have an ancient role in reproductive behavior. Science 338, 540–543.
Ghio, M., Locatelli, M., Tettamanti, A., Perani, D., Gatti, R., Tettamanti, M., 2018. Cognitive training with action-related verbs induces neural plasticity in the action representation system as assessed by gray matter brain morphometry. Neuropsychologia 114, 186–194.
Graybiel, A.M., 1991. Basal ganglia–input, neural activity, and relation to the cortex. Curr. Opin. Neurobiol. 1, 644–651.
Graybiel, A.M., 1995. The basal ganglia. Trends Neurosci. 18, 60–62.
Gregory, S.G., Connelly, J.J., Towers, A.J., Johnson, J., Biscocho, D., et al., 2009. Genomic and epigenetic evidence for oxytocin receptor deficiency in autism. BMC Med. 7, 62.
Haber, S.N., Fudge, J.L., McFarland, N.R., 2000. Striatonigrostriatal pathways in primates form an ascending spiral from the shell to the dorsolateral striatum. J. Neurosci. 20, 2369–2382.
Hikosaka, O., Kim, H.F., Amita, H., Yasuda, M., Isoda, M., et al., 2019. Direct and indirect pathways for choosing objects and actions. Eur. J. Neurosci. 49, 637–645.
Hung, L.W., Neuner, S., Polepalli, J.S., Beier, K.T., Wright, M., Walsh, J.J., Lewis, E.M., Luo, L., Deisseroth, K., Dölen, G., 2017. Gating of social reward by oxytocin in the ventral tegmental area. Science 357 (6358), 1406–1411. https://doi.org/10.1126/science.aan4994.
Johnson, Z.V., Young, L.J., 2017. Oxytocin and vasopressin neural networks: implications for social behavioral diversity and translational neuroscience. Neurosci. Biobehav. Rev. 76, 87–98.
Jurek, B., Neumann, I.D., 2018. The oxytocin receptor: from intracellular signaling to behavior. Physiol. Rev. 98, 1805–1908.
Kim, H.F., Amita, H., Hikosaka, O., 2017. Indirect pathway of caudal basal ganglia for rejection of valueless visual objects. Neuron 94, 920–930-e3.
Kravitz, A.V., Freeze, B.S., Parker, P.R., Kay, K., Thwin, M.T., et al., 2010. Regulation of parkinsonian motor behaviours by optogenetic control of basal ganglia circuitry. Nature 466, 622–626.
Kravitz, A.V., Owen, S.F., Kreitzer, A.C., 2013. Optogenetic identification of striatal projection neuron subtypes during in vivo recordings. Brain Res. 1511, 21–32.
Ladd-Acosta, C., Hansen, K.D., Briem, E., Fallin, M.D., Kaufmann, W.E., Feinberg, A.P., 2014. Common DNA methylation alterations in multiple brain regions in autism. Mol. Psychiatry 19, 862–871.
Lettvin, J.Y., Maturana, H.R., McCulloch, W.S., Pitts, W.H., 1959. What the frog's eye tells the frog's brain. Proc. Inst. Radio Eng. 47, 1940–1951.
Li, K., Nakajima, M., Ibanez-Tallon, I., Heintz, N., 2016. A cortical circuit for sexually dimorphic oxytocin-dependent anxiety behaviors. Cell 167, 60–72-e11.
Llinas, R.R., Roy, S., 2009. The 'prediction imperative' as the basis for self-awareness. Philos. Trans. R. Soc. Lond. Ser. B Biol. Sci. 364, 1301–1307.
Loup, F., Tribollet, E., Dubois-Dauphin, M., Dreifuss, J.J., 1991. Localization of high-affinity binding sites for oxytocin and vasopressin in the human brain. An autoradiographic study. Brain Res. 555, 220–232.

Low, K.L., Ma, C., Soma, K.K., 2017. Tyramide signal amplification permits immunohistochemical analyses of androgen receptors in the rat prefrontal cortex. J. Histochem. Cytochem. 65, 295–308.

Marrocco, J., McEwen, B.S., 2016. Sex in the brain: hormones and sex differences. Dialogues Clin. Neurosci. 18, 373–383.

McCarthy, M.M., Chung, S.R., Ogawa, S., Kow, L.M., Pfaff, D.W., 1991. Behavioral effects of oxytocin: is there a unifying principle? In: Jard, S., Jamison, R.L. (Eds.), Proceedings of the Third International Vasopressin Conference. Editons INSERM; John Libbey Eurotext, Paris, pp. 195–212.

Messinger, D.S., et al., 2015. Early sex differences are not autism-specific: a baby siblings research consortium (BSRC) study. Mol. Autism 6, 32.

Mitre, M., Marlin, B.J., Schiavo, J.K., Morina, E., Norden, S.E., et al., 2016. A distributed network for social cognition enriched for oxytocin receptors. J. Neurosci. 36, 2517–2535.

Moeller, K., Fischer, U., Link, T., Wasner, M., Huber, S., et al., 2012. Learning and development of embodied numerosity. Cogn. Process. 13 (Suppl. 1), S271–S274.

Mullet, E., Cretenet, J., Dru, V., 2014. Motor influences on judgment: motor and cognitive integration. Br. J. Psychol. 105, 69–91.

Nakajima, M., Gorlich, A., Heintz, N., 2014. Oxytocin modulates female sociosexual behavior through a specific class of prefrontal cortical interneurons. Cell 159, 295–305.

Nardone, S., Sams, D.S., Reuveni, E., Getselter, D., Oron, O., Karpuj, M., Elliott, E., 2014. DNA methylation analysis of the autistic brain reveals multiple dysregulated biological pathways. Transl. Psychiatry 4, e433.

Oettl, L.L., Ravi, N., Schneider, M., Scheller, M.F., Schneider, P., et al., 2016. Oxytocin enhances social recognition by modulating cortical control of early olfactory processing. Neuron 90, 609–621.

Parikshak, N.N., Swarup, V., Belgard, T.G., Irimia, M., Ramaswami, G., et al., 2016. Genome-wide changes in lncRNA, splicing, and regional gene expression patterns in autism. Nature 540, 423–427.

Parker, J.G., Marshall, J.D., Ahanonu, B., Wu, Y.W., Kim, T.H., et al., 2018. Diametric neural ensemble dynamics in parkinsonian and dyskinetic states. Nature 557, 177–182.

Peca, J., Feliciano, C., Ting, J.T., Wang, W.T., Wells, M.F., et al., 2011. Shank3 mutant mice display autistic-like behaviours and striatal dysfunction. Nature 472, 437–U534.

Pfaff, D.W., 2017. How the Vertebrate Brain Regulates Behavior: Direct from the Lab. Harvard University Press, Cambridge, MA. 258 p.

Pfaff, D.W., Barbas, H., 2019. Mechanisms for the approach/avoidance decision applied to autism. Trends Neurosci. 42 (7), 448–457.

Pfaff, D., Keiner, M., 1973. Atlas of estradiol-concentrating cells in the central nervous system of the female rat. J. Comp. Neurol. 151, 121–158.

Pfaff, D.W., Rapin, I., Goldman, S., 2011. Male predominance in autism: neuroendocrine influences on arousal and social anxiety. Autism Res. 4, 163–176.

Portmann, T., Yang, M., Mao, R., Panagiotakos, G., Ellegood, J., et al., 2014. Behavioral abnormalities and circuit defects in the basal ganglia of a mouse model of 16p11.2 deletion syndrome. Cell Rep. 7, 1077–1092.

Price, T.F., Harmon-Jones, E., 2010. The effect of embodied emotive states on cognitive categorization. Emotion 10, 934–938.

Ramaswami, G., Geschwind, D.H., 2018. Genetics of autism spectrum disorder. Handb. Clin. Neurol. 147, 321–329.

Sabari, B.R., Zhang, D., Allis, C.D., Zhao, Y., 2017. Metabolic regulation of gene expression through histone acylations. Nat. Rev. Mol. Cell Biol. 18 (2), 90–101.

Sanders, S.J., et al., 2011. Multiple recurrent de novo CNVs, including duplications of the 7q11.23 Williams syndrome region are strongly associated with autism. Neuron 70, 863–885.

Schaafsma, S.M., et al., 2017a. Sex-specific gene–environment interactions underlying ASD-like behaviors. PNAS 114 (6), 1383–1388.
Schaafsma, S.M., Pfaff, D.W., 2014. Etiologies underlying sex differences in autism spectrum disorders. Front. Neuroendocrinol. 35 (3), 255–271.
Selemon, L.D., Goldman-Rakic, P.S., 1985. Longitudinal topography and interdigitation of corticostriatal projections in the rhesus monkey. J. Neurosci. 5, 776–794.
Sestan, N., State, M.W., 2018. Lost in translation: traversing the complex path from genomics to therapeutics in autism spectrum disorder. Neuron 100 (2), 406–423.
Sliwa, J., Freiwald, W.A., 2017. A dedicated network for social interaction processing in the primate brain. Science 356, 745–749.
Subramanian, K., Brandenburg, C., Orsati, F., Soghomonian, J.J., Hussman, J.P., Blatt, G.J., 2017. Basal ganglia and autism—a translational perspective. Autism Res. 10, 1751–1775.
Torres, N., Martins, D., Santos, A.J., Prata, D., Verissimo, M., 2018. How do hypothalamic nonapeptides shape youth's sociality? A systematic review on oxytocin, vasopressin and human socio-emotional development. Neurosci. Biobehav. Rev. 90, 309–331.
Tsunemoto, R.K., Eade, K.T., Blanchard, J.W., Baldwin, K.K., 2015. Forward engineering neuronal diversity using direct reprogramming. EMBO J. 34, 1445–1455.
Turner, T.N., Eichler, E.E., 2019. The role of de novo noncoding regulatory mutations in neurodevelopmental disorders. Trends Neurosci. 42 (2), 115–127.
Wang, W., Li, C., Chen, Q., van der Goes, M.S., Hawrot, J., et al., 2017. Striatopallidal dysfunction underlies repetitive behavior in Shank3-deficient model of autism. J. Clin. Invest. 127, 1978–1990.
Werling, D.M., et al., 2018. An analytical framework for whole genome sequence association studies and its implications for autism spectrum disorder. Nat. Genet. 50, 727–736.
Westberry, J.M., Wilson, M.E., 2012. Regulation of estrogen receptor alpha gene expression in the mouse prefrontal cortex during early postnatal development. Neurogenetics 13, 159–167.
Willsey, A.J., et al., 2013. Coexpression networks implicate human midfetal deep cortical projection neurons in the pathogenesis of autism. Cell 155, 997–1007.
Yeterian, E.H., Pandya, D.N., 1991. Prefrontostriatal connections in relation to cortical architectonic organization in rhesus monkeys. J. Comp. Neurol. 312, 43–67.
Zikopoulos, B., Barbas, H., 2010. Changes in prefrontal axons may disrupt the network in autism. J. Neurosci. 30, 14595–14609.
Zikopoulos, B., Garcia-Cabezas, M.A., Barbas, H., 2018a. Parallel trends in cortical grey and white matter architecture and connections in primates allow fine study of pathways in humans and reveal network disruptions in autism. PLoS Biol. 16.
Zikopoulos, B., Hoistad, M., John, Y., Barbas, H., 2017. Posterior orbitofrontal and anterior cingulate pathways to the amygdala target inhibitory and excitatory systems with opposite functions. J. Neurosci. 37, 14.
Zikopoulos, B., Liu, X., Tepe, J., Trutzer, I., John, Y.J., Barbas, H., 2018b. Opposite development of short- and long-range anterior cingulate pathways in autism. Acta Neuropathol. 135, 759–778.

Further reading

American Psychiatric Association. American Psychiatric Association. DSM-5 Task Force, 2013. Diagnostic and Statistical Manual of Mental Disorders: DSM-5. American Psychiatric Association, Washington, DC. xliv, 947 p.
Barbas, H., 1995. Anatomic basis of cognitive-emotional interactions in the primate prefrontal cortex. Neurosci. Biobehav. Rev. 19, 499–510.

Barbas, H., 2016. Frontal cortex. In: Neuroscience in the 21st Century. Springer, Heidelberg, pp. 1421–1469.

Benowitz, L.I., Routtenberg, A., 1987. A membrane phosphoprotein associated with neural development, axonal regeneration, phospholipid metabolism, and synaptic plasticity. Trends Neurosci. 10, 527–532.

Brick, K., Thibault-Sennett, S., Smagulova, F., Lam, K.G., Pu, Y., et al., 2018. Extensive sex differences at the initiation of genetic recombination. Nature 561, 338–342.

Bullock, D., 2016. Dopamine and its actions in the basal ganglia system. In: Soghomonian, J.J. (Ed.), The Basal Ganglia, Innovations in Cognitive Neuroscience. Springer International Publishing, Switzerland (Chapter 5).

Carrier, N., Kabbaj, M., 2012. Sex differences in social interaction behaviors in rats are mediated by extracellular signal-regulated kinase 2 expression in the medial prefrontal cortex. Neuroscience 212, 86–92.

Clark, A.S., MacLusky, N.J., Goldman-Rakic, P.S., 1988. Androgen binding and metabolism in the cerebral cortex of the developing rhesus monkey. Endocrinology 123, 932–940.

Damasio, A.R., 1999. The Feeling of What Happens: Body and Emotion in the Making of Consciousness. Harcourt Brace, New York. xii, 386 p.

Darnell, J.C., Fraser, C.E., Mostovetsky, O., Stefani, G., Jones, T.A., Eddy, S.R., Darnell, R.B., 2000. Kissing complex RNAs mediate interaction between the fragile-X mental retardation protein KH2 domain and brain polyribosomes. Genes Dev. 19 (8), 903–918.

Delli Carri, A., Onorati, M., Lelos, M.J., Castiglioni, V., Faedo, A., et al., 2013. Developmentally coordinated extrinsic signals drive human pluripotent stem cell differentiation toward authentic DARPP-32+ medium-sized spiny neurons. Development 140, 301–312.

Dhakar, M.B., Stevenson, E.L., Caldwell, H.K., 2013. Oxytocin, vasopressin, and their interplay with gonadal steroids. In: Choleris, E., Pfaff, D.W., Kavaliers, M. (Eds.), Oxytocin, Vasopressin, and Related Peptides in the Regulation of Behavior. Cambridge University Press, Cambridge, pp. 3–26.

Duclot, F., Kabbaj, M., 2015. The estrous cycle surpasses sex differences in regulating the transcriptome in the rat medial prefrontal cortex and reveals an underlying role of early growth response 1. Genome Biol. 16, 256.

Elagoz-Yuksel, M., et al., 2016. The altered promoter methylation of oxytocin receptor gene in autism. J. Neurogenet. 30 (3–4), 280–284.

Flechsig, P., 1901. Developmental (myelogenetic) localisation of the cerebral cortex in the human subject. Lancet 158, 1027–1029.

Gandal, M.J., Zhang, P., Hadjimichael, E., Walker, R.L., Chen, C., et al., 2018. Transcriptome-wide isoform-level dysregulation in ASD, schizophrenia, and bipolar disorder. Science 362.

García-Cabezas, M.A., Zikopoulos, B., Barbas, H., 2019. The structural model: a theory linking connections, plasticity, pathology, development and evolution of the cerebral cortex. Brain Struct. Funct. https://doi.org/10.1007/s00429-019-01841-9.

Geier, D.A., Kern, J.K., Sykes, L.K., Geier, M.R., 2016. Examining genotypic variation in autism spectrum disorder and its relationship to parental age and phenotype. Appl. Clin. Genet. 9, 121–129.

Geschwind, D.H., State, M.W., 2015. Gene hunting in autism spectrum disorder: on the path to precision medicine. Lancet Neurol. 14, 1109–1120.

Goldman, S., 2013. Opinion: sex, gender and the diagnosis of autism—a biosocial view of the male preponderance. Res. Autism Spectr. Disord. 7 (6), 675–679.

Ghashghaei, H.T., Hilgetag, C.C., Barbas, H., 2007. Sequence of information processing for emotions based on the anatomic dialogue between prefrontal cortex and amygdala. NeuroImage 34, 905–923.

Hammock, E.A., Young, L.J., 2006. Oxytocin, vasopressin and pair bonding: implications for autism. Philos. Trans. R. Soc. Lond. Ser. B Biol. Sci. 361 (1476), 2187–2198.

Hanamsagar, R., Bilbo, S.D., 2017. Environment matters: microglia function and dysfunction in a changing world. Curr. Opin. Neurobiol. 47, 146–155.
Hikosaka, O., Ghazizadeh, A., Griggs, W., Amita, H., 2018. Parallel basal ganglia circuits for decision making. J. Neural Transm. (Vienna) 125, 515–529.
Jones, E.G., 2007. The Thalamus. Cambridge University Press, New York.
Kapfhammer, J.P., Schwab, M.E., 1994. Inverse patterns of myelination and GAP-43 expression in the adult CNS: neurite growth inhibitors as regulators of neuronal plasticity? J. Comp. Neurol. 340, 194–206.
Keverne, E.B., 2015. Genomic imprinting, action, and interaction of maternal and fetal genomes. Proc. Natl. Acad. Sci. U. S. A. 112 (22), 6834–6840.
Keverne, B., 2017. Beyond Sex Differences: Genes, Brains and Matrilineal Evolution. Cambridge University Press, Cambridge, United Kingdom; New York, NY. xiii, 217 p.
Kopec, A.M., Fiorentino, M., Bilbo, S.D., 2018. Gut-immune-brain dysfunction in autism: importance of sex. Brain Res. 1693 (2018), 214–217.
Lidow, M.S., Goldman-Rakic, P.S., Gallager, D.W., Rakic, P., 1991. Distribution of dopaminergic receptors in the primate cerebral cortex: quantitative autoradiographic analysis using [3H]raclopride, [3H]spiperone and [3H]SCH23390. Neuroscience 40, 657–671.
Marlin, B.J., Mitre, M., D'Amour, J.A., Chao, M.V., Froemke, R.C., 2015. Oxytocin enables maternal behavior by balancing cortical inhibition. Nature 520, 499–504.
Murugan, M., Jang, H.J., Park, M., Miller, E.M., Cox, J., et al., 2017. Combined social and spatial coding in a descending projection from the prefrontal cortex. Cell 171, 1663–77 e16.
Nakamura, K., Hikosaka, O., 2006. Role of dopamine in the primate caudate nucleus in reward modulation of saccades. J. Neurosci. 26, 5360–5369.
Neumann, I.D., Slattery, D.A., 2016. Oxytocin in general anxiety and social fear: a translational approach. Biol. Psychiatry 79, 213–221.
Noritake, A., Ninomiya, T., Isoda, M., 2018. Social reward monitoring and valuation in the macaque brain. Nat. Neurosci. 21, 1452–1462.
Onorati, M., Castiglioni, V., Biasci, D., Cesana, E., Menon, R., et al., 2014. Molecular and functional definition of the developing human striatum. Nat. Neurosci. 17, 1804–1815.
Pfaff, D.W., Rubin, R., Schneider, J., Head, G., 2018. Principles of Hormone/Behavior Relations. Elsevier Science & Technology, New York, NY.
Rakic, P., 2002. Neurogenesis in adult primate neocortex: an evaluation of the evidence. Nat. Rev. Neurosci. 3, 65–71.
Sanders, S.J., 2019. Next-generation sequencing in autism Spectrum disorder. Cold Spring Harb. Perspect Med. 9 (8).
Sanders, S.J., et al., 2012. De novo mutations revealed by whole-exome sequencing are strongly associated with autism. Nature 485, 237–241.
Sanders, S.J., et al., 2015. Insights into autism Spectrum disorder genomic architecture and biology from 71 risk loci. Neuron 87, 1215–1233.
Sato, D., et al., 2012. SHANK1 deletions in males with autism spectrum disorder. Am. J. Hum. Genet. 90 (5), 879–887.
Schaafsma, S.M., Gagnidze, K., Reyes, A., Norstedt, N., Mansson, K., et al., 2017b. Sex-specific gene-environment interactions underlying ASD-like behaviors. Proc. Natl. Acad. Sci. U. S. A. 114, 1383–1388.
Sinha, P., Kjelgaard, M.M., Gandhi, T.K., Tsourides, K., Cardinaux, A.L., et al., 2014. Autism as a disorder of prediction. Proc. Natl. Acad. Sci. U. S. A. 111, 15220–15225.
Siu, M.T., Weksberg, R., 2017. Epigenetics of autism spectrum disorder. Adv. Exp. Med. Biol. 978, 63–90.
Victor, M.B., Richner, M., Hermanstyne, T.O., Ransdell, J.L., Sobieski, C., et al., 2014. Generation of human striatal neurons by microRNA-dependent direct conversion of fibroblasts. Neuron 84, 311–323.

Victor, M.B., Richner, M., Olsen, H.E., Lee, S.W., Monteys, A.M., et al., 2018. Striatal neurons directly converted from Huntington's disease patient fibroblasts recapitulate age-associated disease phenotypes. Nat. Neurosci. 21, 341–352.
Wichmann, T., 2018. Models of Parkinson's disease revisited. Nature 557, 169–170.
Yeterian, E.H., Pandya, D.N., 1995. Corticostriatal connections of extrastriate visual areas in rhesus monkeys. J. Comp. Neurol. 352, 436–457.

CHAPTER 10

Loss

This chapter treats a reason completely different from autism for loss of social contact. Autism involves children (Chapter 2) who, for internal psychological reasons, cannot help but avoid contact. These children (mostly male) are on the autism spectrum, and neurobiologists understand their motivation fairly well (but not entirely). Forebrain pathways involved in this form of social avoidance were just described (Chapter 9).

A completely different reason for loss of social contact is that a person becomes lonely due to *external* circumstances, again because of circumstances beyond their control.

Urgently, in an age of coronaviruses when people are often forced into seclusion, thinking about the neurobiology of loneliness must be brought forward.

Loneliness is a major problem of post-industrial society (42% of Americans live alone, 61% under the age of 35), but it is not essentially a problem that neurobiologists can functionally address. It can best be addressed by altering an individual's social circumstances, although its potential symptoms (e.g., depression, anxiety) are amenable to treatment by means of mood-altering drugs.

Loneliness

Most of us would claim that we know what loneliness is, probably firsthand. But what is it, really? Clearly, it is not the occasional "solitude" referred to by Benjamin Franklin as a respite from over-activity. Such solitude provides an opportunity to focus one's mind on something important—perhaps even to be introspective. It allows people like writers (and, I might add, scientists) to be productive without feeling scattered. Finally, there is space to get organized. At times, we have all sought solitude, especially in crowded environments where too many people are making too many demands all at the same time. Solitude has positive valences, and it is within a person's control. We can step out of it when we've had enough.

Origins of Human Socialization
https://doi.org/10.1016/B978-0-323-85861-8.00001-3

On the other hand, loneliness is a disquieting emotional response to isolation. Loneliness is being disconnected from others (or a significant other) against one's will and, most likely, without an end in sight. The same conditions that support solitude (e.g., an empty house) can produce loneliness, but with markedly different effects and a different response. In response to loneliness a person seeks to be reconnected to an accustomed network (or, in a pinch, to any network that may alleviate the isolation) or to a specific person to whom they have grown attached.

Loneliness has a vast array of causes, not least a lack of communication characteristic of places with low population densities. In these situations, loneliness can feed on itself—the instinct to reach out to and connect with others is continually short-circuited; the insuperable, apparently chronic nature of social disconnection can drive people to withdraw even further. In these situations, loneliness can lead to depression, suicidal thoughts, addiction, and the total collapse of hope. It is hard for people so afflicted to imagine the future except as the dull extension of an endless, intolerable present. With no one around to take an interest in your self-expression, why even bother?

Loneliness is evidence that humans need communication with other members of their species, and that the social impetus that I have been discussing is crucial to our survival. In a much starker context, "isolation" is one of the greatest punishments that prisons can impose; prisoners readily peel potatoes for pennies an hour rather than sit in a cell alone with a toilet. The term penitentiary derives from the Latin for "repentance," and those consigned to penitentiaries were locked away sometimes indefinitely to contemplate their crimes.

The subject of loneliness has been studied at length. About 50 years ago, it was anatomized into social loneliness (resulting from the lack of a social network) and its emotional counterpart (the absence of deep, nurturing relationships with others (Weiss, 1973). More recently, three types of loneliness were described (Cacioppo and Cacioppo, 2018), relating to the amount of space commanding the subject's attention. "Intimate loneliness" refers to lost love, and involves the closest space surrounding a person. The subject individual suffers from the absence of "a person one can rely on for emotional support during crises, who provides mutual assistance, and who affirms one's value as a person." A second type of loneliness refers to an absence of "quality friendships," including family connections. This social domain relates to a space "delimitated by the multi-modal, personal, communications and interactions." A third type, "collective loneliness," deals with groups, teams, and even national identities. The more voluntary the association, the less it results in collective loneliness. Cacioppo was quoted in *The Guardian* as saying that "Chronic loneliness increases the odds of an early death by twenty percent," but he did not define the type of loneliness and it is likely that is statement did not involve the third category.

Frequently, however, the first two categories (or both in the case of Weiss) segue into the other, since without a social network it is hard to find friends or a romantic partner. The reverse can also be true, i.e., without a romantic partner it is often harder (especially for women) to maintain a social network—widows and divorcees complain that invitations to gatherings are hard to come by, since these women are viewed as "extra" or even as threats to their married friends. Such intangible forms of social stigma are a frequent cause of loneliness in people who never before experienced it and do not know how to cope. Indeed, the issue of loneliness among the elderly—especially among older women, who often outlive their mates—is hugely important especially because as they age, the elderly have fewer friends, cease being able to drive, and are less able to get out of the house on their own. So-called retirement homes offer an array of activities to take people's minds off themselves. Nonetheless, the National Academy of Medicine, as reported by Janet Adamy, stated that a substantial number of older Americans "who live in community settings have few relationships or infrequent social contact."

At Duke University, Blazer and Hybels (2005) reviewed "biological (genetics and heredity factors, neurotransmitter dysfunction, endocrine changes, vascular disorders, and medical co-morbidities), psychological (personality attributes, neuroticism, cognitive distortions, and the lack of emotional control and self-efficacy) and social (stressful life events, bereavement, chronic stress or strain, socio-economic disadvantage and impaired social support) origins of late-life depression." They concluded that "Older adults appear to be at greater risk for major depression biologically, yet the actual frequency of depression is lower compared to younger adults." Possibly, the psychological effects of social selectivity and wisdom offer the elderly some protection, but they are still prone to depression if only because they can be frail and are likelier to become seriously sick more easily than the young.

Loneliness is difficult to measure because it is so subjective. Some people may have a higher tolerance for social detachment than others, and may fear being stigmatized if they admit to having a hard time. Men especially may feel that such an admission signifies weakness. But we do know that some groups suffer more than others—e.g., the poor, the unemployed, the homeless, recent immigrants who do not speak the language—if for no other reason than structural barriers prevent ready socializing with other, more settled individuals. A survey by *The Economist* reported that across countries, roughly 10–20% of people often felt lonely, left-out, or isolated. Different scales in different studies yield different percentages, but it is widely perceived among doctors that loneliness is a public health problem in many countries.

My concern with loneliness is with its potential effects. I have mentioned suicidal thoughts—some of which actually result in suicide, e.g., among college students away from home for the first time, overwhelmed by the relentless demands of labs, classes, and a need for acceptance by their peers. While loneliness cannot be regarded as the sole reason for these tragedies, it is also true that these students felt unable to reach out for help. They were isolated within their own troubled minds. They lacked the necessary skills—and were too dispirited—to slow down, connect with the right people, and seek help in evaluating their options.

But short of suicide, loneliness poses other risks to health and well-being. It increases the risk of stroke and cardiovascular disease (*Pfizer Journal*, 2000), and is linked to increased levels of hypertension, high cholesterol, and obesity, as understood very early by Chicago professor John Cacioppo. John Capitanio and Steve Cole (2019) state that "loneliness in humans confers significant risk for morbidity and mortality, although mechanisms remain unclear. For studies of biological effects, we have to use monkeys." Lonely monkeys (those that display high frequencies of social initiations but low frequencies of complex interaction) show elevated sympathetic nervous system activity and down regulated Type I interferon responses. That means an immunological deficiency.

In Mainz, Germany, Xia and Li (2018) agree that "Population-based studies have demonstrated that both objective social isolation and the perception of social isolation (loneliness) are correlated with a higher risk of mortality and that both are clearly risk factors for cardiovascular disease (CVD). Lonely individuals have increased peripheral vascular resistance and elevated blood pressure. Socially isolated animals develop more atherosclerosis than those housed in groups." But the mechanisms "responsible for the increased cardiovascular risk are poorly understood. In recent reports, loneliness and social stress were associated with activation of the hypothalamic-pituitary-adrenocortical axis and the sympathetic nervous system. Repeated and chronic social stress leads to glucocorticoid resistance, enhanced myelopoiesis, upregulated proinflammatory gene expression, and oxidative stress." At the very least these mean increased inflammation, which is well known to lead to increased chances of heart attacks.

In England, Valtorta et al. (2018) at York and Newcastle also cite "increasing evidence of an association between social relationships and morbidity in general, and cardiovascular disease in particular." By a clever experimental design that allowed them to separate inferences about the causative effects of loneliness as opposed to social isolation (even though they often occur together), Valtorta was able to conclude that "loneliness is associated with an increased risk of developing coronary heart disease and stroke, independently of traditional cardiovascular disease risk factors" as opposed to social isolation without loneliness.

A huge theme in this area of scientific work was opened up by geneticist and immunologist Steve Cole, working together with the Cacioppo team (2015a,b) "to define the cellular mechanisms of up-regulated inflammatory gene expression and down-regulated antiviral response in people experiencing perceived social isolation (loneliness), they conducted integrative analyses of leukocyte (white blood cell) gene regulation in humans and rhesus monkeys." They reported that older adults showed up-regulation of the sympathetic nervous system, monocyte (white blood cell) population expansion, and up-regulation of blood cell gene transcriptional response to adversity (CTRA). To determine underlying mechanisms, follow-up work with monkeys identified genetic, hormonal and immunological abnormalities due to social isolation.

Continuing on this theme, Chun et al. (2017) followed up on previous reports which proved up-regulation of proinflammatory gene expression and down-regulation of Type I interferon anti-viral genes in circulating blood cells. What about "social instability in adult male rhesus macaques"? Amazingly, unstable social conditions down-regulated axillary blood cell expression of genes involved in Type I interferon anti-viral responses. Genetic transcript origin analyses implicated specific types of white blood cells as cellular mediators of these effects, and genetic promoter-based bioinformatics analyses indicated reduced activity of specific stress-related gene transcription factors. These pioneering genomic studies will lead to a new generation of mechanistic studies on loneliness and social isolation.

Many claims have been made that social media and e-gaming can cause isolation and loneliness if used immoderately. See, e.g., the Harvard Health Blog (Dec. 21, 2018), which compares overuse of social media to overeating—both are bad for your health. Nowland et al. (2018) take a similarly nuanced position, recognizing that it is too soon for broad and sweeping conclusions. They currently suggest this wait-and-see approach: "When the Internet is used as a way station on the route to enhancing existing relationships and forging new social connections, it is a useful tool for reducing loneliness. But when social technologies are used to escape the social world and withdraw from the 'social pain' of interaction, feelings of loneliness are increased." One cannot help but worry when we read about the pathologies of social media on a weekly basis. In the last chapter of this book, I will discuss some of the concerns regarding this topic. Because virtual on-line communities are populated by "transients" with no commitment to anyone else in the community (and who suffer no adverse consequences should they leave), vulnerable individual participants may be taking psychological risks if they become deeply involved.

Neuroendocrine consequences of loneliness. As I have indicated, loneliness can be lethal. In one dramatic demonstration, Jennifer Tung et al. (2016), an anthropologist at Duke University, studied wild female baboons and showed that individuals who had suffered early adversity were socially isolated in adulthood and died ten or more years younger than females not so isolated. Short of ultimate effects on longevity, the Cacioppo team has described a host of problems due to loneliness: depression, social anxiety, high blood pressure, sleep problems, immunological disorders and eventual cognitive decline.

Loneliness increases the body's concentration of cortisol, resulting is depression, digestive issues, and the inability to sleep. It is associated with impaired cellular immunity (Cacioppo, 2010), which can reduce the effectiveness of vaccines. There is evidence suggesting that loneliness increases the progression of HIV in HIV-positive men (Cacioppo, 2010). And there's more. Loneliness can impair cognition and willpower and increase the inflammatory response to stress. The National Academy of Medicine study, cited above, stated that "social isolation is associated with a significantly increased risk for early death from all causes."

Thinking about a "neurology of loneliness" (Cacioppo et al., 2014) remains at this point a primarily aspirational endeavor. The subject is inherently hard to study because data regarding the lonely person often stays private, and there is too much of a gulf between animal research and clinical projects. Nevertheless, studies have claimed unusual fMRI activation of neurons in the basal forebrain in response to pleasant objects rather than pleasant people. Also, consider abnormalities in activation of neurons in the temporoparietal junction, a cortical region devoted, in part, to enabling social astuteness including the understanding of the intentions of other people. It has been suggested that endocrine mechanisms are also likely involved.

For example, Cacioppo et al. (2015a,b) showed that chronic social isolation increases the activation of the hypothalamic pituitary adrenocortical axis. Hawkley et al. (2012) reviewed the effects of social isolation on glucocorticoid (e.g., cortisone) physiology in social mammals. It was messy: "Evidence indicates that objective and perceived social isolation alter HPA regulation, although the nature and direction of the HPA response differs among species and across age." For the purpose of this book's arguments, the absence of animal models useful for drawing conclusions about social isolation in humans was particularly telling: it explained the difficulty of medical progress in this field.

Recently, though, studies of the consequences of social isolation have taken a molecular and genetic turn, notably in the laboratory of David J. Anderson, at Cal Tech (Zelikowsky et al., 2018). These studies focused on tachykinins, neuropeptides well known for exciting nerve cell activity and making smooth muscles contract. The gene Tac2 produces the neuropeptide neurokinin B. In their new Cell paper, Anderson and his colleagues reported that stress from social isolation causes significant elevations of neurokinin B levels "throughout the brain" and suggest that such social isolation institutes a change in "brain state." Administering a neurokinin receptor blocker prevented behavioral effects of social isolation, while elevating neurokinin B expression caused some of the behavioral effects of social isolation even when the mice were not socially isolated. These results up the ante in social isolation analyses because different circuitries related to different brain regions now must be taken into account.

At Duke University, genetics and evolutionary anthropology professor Jenny Tung looked at social environmental effects on gene regulation (Tung and Gilad, 2013). With respect to studies of social environmental effects in humans and non-human primates, their review focused on genome-wide studies in peripheral blood mononuclear cells. For monkeys, a frequently used paradigm has been "...rhesus macaques reared with their mothers under normal social conditions were compared to individuals reared either with same-aged peers or by 'surrogates' (terry cloth-covered hot water bottles), with short daily periods of peer exposure." Under these circumstances several hundred transcripts showed "two-fold (or higher) differences in gene expression differentiated the mother-reared (control) animals from those that experienced substantial early life adversity." Yes, strikingly, absence of mothers led to changes of gene expression in the brain.

Snyder-Mackler and Tung (2014) compared aging effects and effects of chronic social stress on gene regulation in blood cells and found the direction of significant gene expression changes tends to be shared between the two data sets. "This overlap was observable at three levels: (i) individual genes; (ii) general functional categories of genes; and (iii) molecular pathways implicated in aging." Especially important are gene expression changes for hormonal responses to stress (Snyder-Mackler et al., 2019). Steroid stress hormones such as cortisol have effects on gene expression mediated by their binding to a protein called the glucocorticoid receptor, a ligand-activated gene transcription factor. But Snyder Mackler and Tung raise the possibility that access to DNA by the glucocorticoid receptor may itself be regulated by a monkey's social environment. They studied immune cell gene regulation in female rhesus macaques, using paired control and cortisol-treated peripheral blood mononuclear cell samples. Social status "not only influences immune cell gene expression but also chromatin (covering DNA) accessibility at hundreds of regions in the genome." These effects of social status in a primate on chromatin changes which regulate DNA accessibility are especially exciting because they amount to an example of what we now call "epigenetic" effects: chemically well-defined effects on the regulation of the N-termini of histone proteins which comprise the "chromatin" coating of DNA.

Thus, social isolation has consequences in primates and, likely humans, they can now be studied with the newest genetic and epigenetic techniques.

About therapeutic approaches, the science writer Lydia Denworth (2020) deals well with recent work on loneliness when she writes about friendship which, she states "serves the elemental need to belong and triggers physical pleasure and pain to make us pay attention to that need" (p. 4). She suggests that "the striking thing about loneliness is not that those who suffer are peculiar, but that they are so ordinary." (p. 77) In a large poll, about half of Americans "reported sometimes or always feeling alone." It could be due to the simple absence of social contact or a "lack of emotional closeness" (p. 78).

There are also specific approaches with regard to the elderly. David Blazer (2002) faced the fact that "feelings of sadness and loneliness are ubiquitous in late life and are a risk factor for depression and perhaps other mental illnesses in late life. Targeting sadness and loneliness for an intervention would address both primary risk reduction for depressive disorders and promotion of overall mental health in the elderly." Blazer argues that "the attainment of positive mental health depends in considerable part upon an individual's self-efficacy -- the belief that one can organize and execute the courses of action required to develop and enhance a person's belief that he or she can act in ways that lead to a desired goal." He advocates for programs to enhance self-efficacy by arranging for "the experience of successfully dealing with and thus overcoming specific problems." Such programs could be made available to individuals but also, indeed, to entire communities.

The point is that loneliness has finally become recognized as a major social issue. Two hundred years ago, most people lived on farms. They worked together and ate together. Now we live in apartment buildings where the population is transient, and we work in cubicles or remotely from home. Precisely because addressing the condition of loneliness speaks to the living conditions of large numbers of people, scientists like the late University of Chicago professor John T. Cacioppo, Stephanie Cacioppo and their lab (2018) "emphasized the importance of improving health care and easing the pain of those suffering from emotional distress syndromes including loneliness." They estimated that the percentage of Americans suffering from loneliness has more than doubled during the past 40 years, and quoted similar results from England, New Zealand and China.

In their scientific work the Cacioppo team showed how solitariness can be addressed through sharing, working together, and the formation of groups dedicated to solving a massive problem. It is interesting to speculate whether the three categories of loneliness addressed by their work derive in part from their personal experience (loss of love, because they found love; loss of quality friendships, because that is what they formed as colleagues; and the loss of a larger group of colleagues, since they helped build a small industry of people studying aspects of loneliness).

About the present Covid-19 situation in many countries, it could be said that loneliness is an epiphenomenon of quarantine. Even people who did not have the virus were told to keep away from each other, let alone from people who might be contagious. It is difficult to overstate the toxic mix of loneliness, stress, and fear that gripped thousands of people as the pandemic lumbered its way into virtually every community. People who were proud of being able to live on their own, especially the elderly in this category, suddenly felt vulnerable. As I wrote this chapter, at the height of the pandemic in New York, it was easy to see how loneliness could almost overnight transition into an existential challenge. Who knows whether our technological fixes to loneliness will become permanent fixtures of a population always a little more frightened that anyone might care to admit? As I have speculated throughout this book, technology may be poised to carry the evolution of (wo)man/machine systems into uncharted territories for which we may be maladapted.

The next chapter concerns genetics and epigenetics. Gene expression permits and even encourages some forms of social behavior, while gene transcriptional effects of hormones affect brain circuits relevant for social behavior. Epigenetics refers to means by which gene expression is changed without changes in DNA itself. Damage to certain genes or to epigenetic mechanisms regulating those genes can negatively affect social behavior. The point is that the human instinct for social engagement, though "wired" into our brains, is still sufficiently delicate that it can be short-circuited.

New questions

1. How will the complexities of immunology be understood as they contribute to the neurobiological consequences of loneliness?
2. What will be the long-term social consequences of the quarantines required by coronaviruses?
3. Literary and philosophical explorations of loneliness have been legion. How will modern neuroscientific scholarship affect what some authors and philosophers say?

References

Blazer, D.G., 2002. Self-efficacy and depression in late life: a primary prevention proposal. Aging Ment. Health 6 (4), 315–324.

Blazer, D.G., Hybels, C.F., 2005. Origins of depression in later life. Psychol. Med. 35 (9), 1241–1252.

Cacioppo, J.T., Cacioppo, S., 2018. The growing problem of loneliness. Lancet 391 (10119), 426–451.

Cacioppo, S., et al., 2014. Toward a neurology of loneliness. Psychol. Bull. 140, 1464–1488.

Cacioppo, J.T., et al., 2015a. The neuroendocrinology of social isolation. Ann. Rev. Psychol. 66, 733–778.

Cacioppo, S., et al., 2015b. Loneliness: clinical import and interventions. Perspect. Psychol. Sci. 10 (2), 238–249.

Capitanio, J.P., Cacioppo, S., Cole, S.W., 2019. Loneliness in monkeys: neuroimmune mechanisms. Curr. Opin. Behav. Sci. 28, 51–57.

Chun, K., et al., 2017. Social regulation of the lymph node transcriptome in rhesus macaques (Macaca mulatta). Psychoneuroendocrinology 76, 107–113.

Denworth, L., 2020. Friendship. WW Norton, New York.

Hawkley, L.C., et al., 2012. Effects of social isolation on glucocorticoid regulation in social mammals. Horm. Behav. 62 (3), 314–323.

Nowland, R., et al., 2018. Loneliness and social internet use: pathways to reconnection in a digital world? Perspect. Psychol. Sci. 13 (1), 70–87.

Snyder-Mackler, N., Somel, M., Tung, J., 2014. Shared signatures of social stress and aging in peripheral blood mononuclear cell gene expression profiles. Aging Cell 13 (5), 954–957.

Snyder-Mackler, N., et al., 2019. Social status alters chromatin accessibility and the gene regulatory response to glucocorticoid stimulation in rhesus macaques. Proc. Natl. Acad. Sci. 116 (4), 1219–1228.

Tung, J., Gilad, Y., 2013. Social environmental effects on gene regulation. Cell. Mol. Life Sci. 70 (22), 4323–4339.

Tung, J., et al., 2016. Cumulative early life adversity predicts longevity in wild baboons. Nat. Commun. 19 (7), 11181.

Valtorta, N.K., et al., 2018. Loneliness, social isolation and risk of cardiovascular disease in the English longitudinal study of ageing. Eur. J. Prev. Cardiol. 25, 1387–1396.

Weiss, R., 1973. Loneliness: The Experience of Emotional and Social Isolation. Norton, New York.

Xia, N., Li, H., 2018. Loneliness, social isolation, and cardiovascular health. Antioxid. Redox Signal. 28, 837–851.

Zelikowsky, M., et al., 2018. The neuropeptide Tac2 controls a distributed brain state induced by chronic social isolation stress. Cell 173 (5), 1265–1279.

Further reading

Baron-Cohen, S., 2002. The extreme male brain theory of autism. Trends Cogn. Sci. 6, 2480254.

Cole, S.W., et al., 2015. Myeloid differentiation architecture of leukocyte transcriptome dynamics in perceived social isolation. Proc. Natl. Acad. Sci. U. S. A. 112 (49), 15142–15147.

Hung, L.W., et al., 2017. Gating of social reward by oxytocin in the ventral tegmental area. Science 357, 1406–1409.

Robinson, E.B., et al., 2013. Examining and interpreting the female protective effect against autistic behavior. Proc. Natl. Acad. Sci. U. S. A. 110 (13), 5258–5262.

Werling, D.M., Geschwind, D.F., 2013. Understanding sex bias in autism spectrum disorder. Proc. Natl. Acad. Sci. 10, 4868–4869.

CHAPTER 11

Genetics and epigenetics

The newest scientific contributions to how we understand our social behaviors come from genetics (coding of DNA itself) and gene transcription (Fig. 1) and epigenetics (chemistry affecting the readout from DNA codes without changing the DNA itself (Fig. 2).

Now I have gone back over all the other chapters to pick up some of the genes mentioned therein, so I can explain a bit of what they are and how they work. In particular, where possible, I'll envision how each gene contributes to the regulation of social behavior.

Spoiler alert. The sections of this chapter do not fit together logically like a well-woven story. That is not how genetic discoveries work. That is not how they feed into modern biology and neurobiology. Instead, the separate sections simply illustrate how some recent genetic findings contribute in their own separate ways, to segments of my overall argument.

Brain growth

In Chapter 1, about our ancient forebearers, I mentioned the NOTCH gene as one important determinant of the numbers of neurons in the mammalian cortex. To begin with, it is almost impossible to overestimate the complexity of NOTCH gene biology because it is associated with so many other genes in the cortex. The gene itself codes for a protein which spans a cortical cell's membrane and acts as a receptor. When a biochemical successfully binds to NOTCH, it twists NOTCH so that an enzyme can cleave out a small portion, and that small portion of NOTCH races to the cell nucleus and, together with another protein stimulates stimulates transcription of target genes (Bray, 2016). According to a review by Hevner (2019), one way NOTCH affects the developing cerebral cortex is to bind to "radial glia (RG) progenitors, the neural stem cells of developing cortex."NOTCH makes these radial glia progenitors remain as neural stem cells, thus allowing for more extensive proliferation of the cortex. Working together with another gene, DELTA, NOTCH encourages radial glia cells

Origins of Human Socialization
https://doi.org/10.1016/B978-0-323-85861-8.00015-3

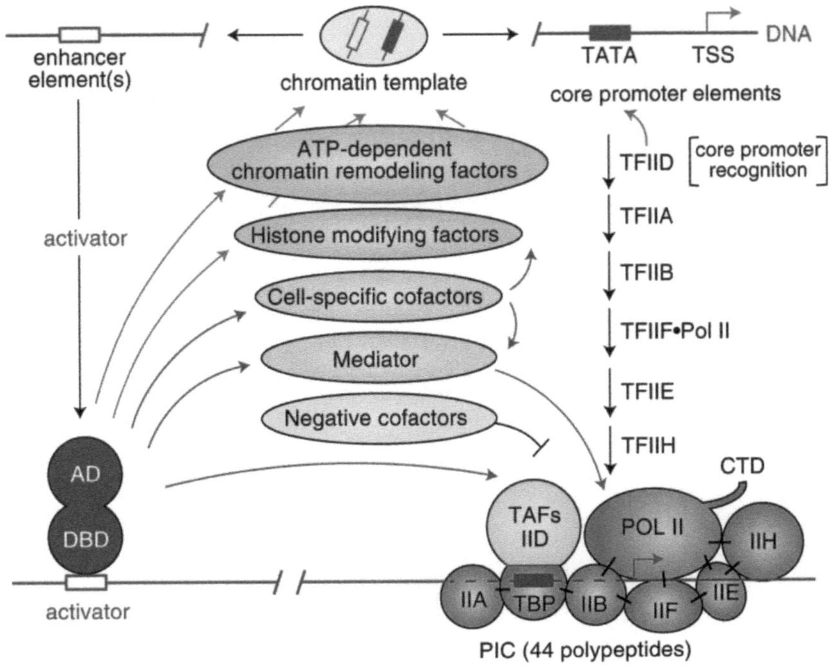

Fig. 1 As pioneered by the work of Robert Roeder, a contemporary sketch of some of the regulations over the initiation of gene transcription. Multiple opportunities for increasing or decreasing probability of transcription are seen. *(Modified from Roeder, R.G., 2019. 50+ Years of eukaryotic transcription: an expanding universe of factors and mechanisms. Nat. Struct. Mol. Biol. 26 (9), 783–791.)*

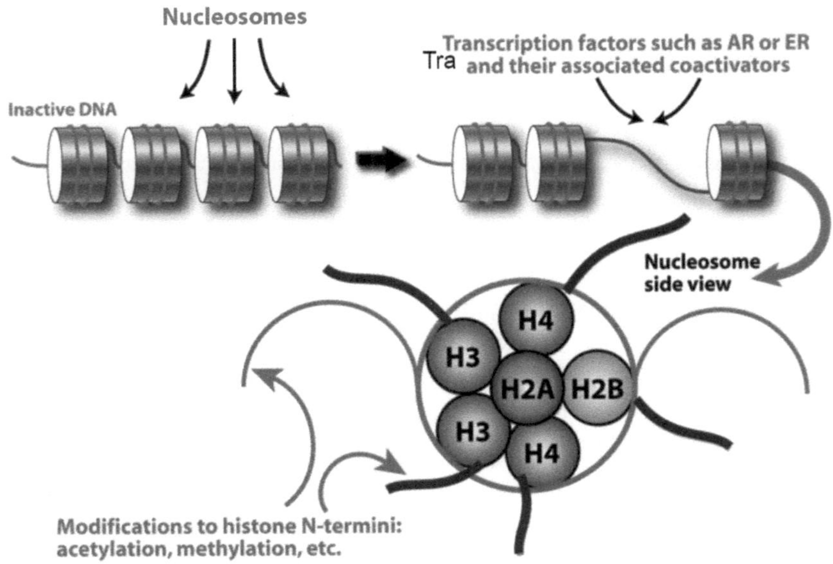

Fig. 2 Inactive chromatin, covering inactive DNA, is bound up in "nucleosomes." Chromatin protein N-termini are available for chemical modifications which, in separate cases, can increase or decrease access by transcription factors (like androgen receptors, AR, or estrogen receptors, ER) to underlying DNA.

to guide interneuron and microglia migration to actually form the cerebral cortex, and they also promote thalamocortical axon growth.

Now, how does more extensive proliferation of the cortex promote social contacts? Well, consider the visual mechanisms, the auditory mechanisms, and the physiological mechanisms underlying altruistic behaviors reviewed in this book. All of those depend on the full growth of the cerebral cortex as we know it.

Without adequate brain size and computing power, we could not manage the prosocial mechanisms I have written about in Chapters 5 and 6. More specifically, some developmentally crucial genes relate to speech, in addition to other functions. FoxP2 is getting famous for this.

Forkhead genes, FoxP2

Another class of genes involved during brain development are the family of forkhead genes, named because mutant embryos had two spiked head structures. Forkhead gene products are transcription factors; they are proteins which, when they bind to specific sequences in DNA, facilitate the transcription of messenger RNAs from specific genes. Forkhead gene-dependent proteins have a so-called "forkhead box" of more than 70 amino acids that can bind to the chromatin overlying DNA even when those chromatin proteins are tightly wound up and seemingly unavailable. More than 100 forkhead genes are subdivided into families according to which tissues they are expressed in, what downstream cascades of genes they trigger, and what functions they are known or supposed to have. Many are involved during embryonic development.

Audition and speech in social communication. I mentioned the forkhead gene FoxP2 during the chapter on evolved auditory mechanisms serving social relations. Not only do FoxP2 mutations cause disorders of human speech, but they also disrupt aspects of vocal learning, as shown by experiments on bird song. In the words of Constance Scharff at the Free University of Berlin (Fisher and Scharff, 2009) writing in *Trends in Genetics*, "Rare mutations of the FOXP2 transcription factor gene cause a monogenic syndrome characterized by impaired speech development and linguistic deficits." Moreover, "reduced FoxP2 dosage yields abnormal synaptic plasticity and impaired motor-skill learning in mice, and disrupts vocal learning in songbirds." In other words, FoxP2 contributes powerfully to capacities for speech, but not because it is specifically devoted to speech. Instead, normal speech development requires motor learning components which depend on FoxP2.

Functional genomic and bioinformatics analyses "of the putative targets of Fox P2 indicated roles in CNS patterning, development and function, including signal transduction, neurite outgrowth and axon guidance, neurotransmission and synaptic plasticity". Interested in discerning genetic relations between different language disorders, Fisher and Scharff highlighted a fascinating finding: studies identified a FOXP2-bound fragment which mapped within intron 1 of the *CNTNAP2*. Yes, FoxP2 helps to regulate language development, and studies primarily by Daniel Geschwind at UCLA (followed up by my lab, [Schaafsma et al., 2017]) clearly implicate CNTNAP2 in autism. Altogether, reading Fisher and Scharff, I am left with the impression that FoxP2 mutation effects on language development and vocal learning constitute a special, and especially interesting case of disrupted motor learning.

In Chapter 4, I summarized a leading field in neurobiology, the science of how the brain manages reproductive biology, especially reproductive behavior. The genes involved have received a lot of attention, as I summarize now.

Sex

How does gene expression in the brain regulate prototypical social behaviors: sex? Genes which regulate reproduction when expressed in brain obviously contribute to the initiation of at least some social behaviors. The "master gene" for stimulating reproductive processes in males and females, lab animals and humans, is called Gonadotropin Releasing Hormone (GnRH). GnRH is a neurochemical which consists of 10 amino acids, and actually comprises a small portion of its much larger gene. GnRH facilitates reproductive behavior (Pfaff, 1973). In genetic terms, how does this work?

There are two GnRH genes in humans. GnRH gene structure, as reviewed by Belsham and Lovejoy (2005) shows that even though the 10 amino acids of GNRH take only 30 nucleotide bases of DNA to code, the two copies of these 30 bases are found in two huge genes, one more than over 4000 nucleotide bases and the other over 2000.

Why two genes? Because the two separate controlling regions of DNA "upstream" in the chromosome allow for different, subtle and physiologically appropriate regulation patterns of gene expression. For example, in neurons, several different transcription factors can bind "upstream" in the so-called promoter and enhancer regions of the gene and the subsequent products of gene expression lead to different biochemical events in the cytoplasm and on the surface of the neuron, events (in the case of GnRH) of importance for the regulation of reproduction.

It turns out that these GnRH gene-expressing neurons are the only neurons born outside the brain. At Rockefeller in the '80s, Marlene Schwanzel-Fukuda and I were completing a developmental study of these of GnRH expression where we could see GnRH cells in nose and in brain and follow the pattern of expression as a function of time. It suddenly struck me that we were looking at an actual *migration* of these neurons, a thought that depended on when I was helping my Ph.D. advisor at MIT trace other migrating neurons in the brain. These GnRH neurons are born in the nose and migrate up the nose, along the bottom of the brain, and take up their functional positions in a portion of the anterior hypothalamus called the preoptic area (Schwanzel-Fukuda and Pfaff, 1989). Over the years, we saw that this GnRH neuron migration was true among all vertebrates, from fish (Parhar et al., 1995, 1996) to humans (Schwanzel-Fukuda et al., 1989, 1996). As I explain below, if these neurons do *not* complete their migration from nose to brain, patients have no libido (Schwanzel-Fukuda et al., 1989).

The genetic basis for what we found started out simple and then got more complicated. As reviewed by Harvard reproductive endocrinologist Stamou and Georgopoulos (2018), the first gene found to be responsible for the migration failure of GnRH neurons, KAL1, "is localized to the distal portion of X chromosome (Xp22.3)."The protein encoded by this *KAL1* gene is associated with neural cell adhesion and axonal migration. So this makes sense, because migrating GnRH neurons would need nerve cell adhesion as well as axonal migration.

Then things got more complicated. For example, Kim et al. (2018) reported that that a different gene, WDR11, involved in powerful signaling pathways during development, is essential for normal ciliogenesis. Cilia are essential for fluid transport and confer motility. Thus, it makes sense that, as GnRH are trying to migrate along ciliated epithelia in the nose and at the nose/brain interface, abnormalities or absence of cilia could bring the GnRH neurons to a halt. And other genes, too numerous to mention here, are involved as well. The story is still developing.

The Kallmann's Syndrome libido-loss provides an interesting example of gene/behavior causal relations in humans, as follows: Kallmann's patients have no libido *because* they have very low testosterone. They have low testosterone *because* of deficient luteinizing hormone (LH) coming in the bloodstream to the testes from the pituitary gland. LH is deficient *because* there is a loss of GnRH in the tiny vessels going from the hypothalamus to the pituitary gland. There is a loss of GnRH there *because* of an absence of GnRH neurons in the hypothalamus. There is an absence of GnRH neurons *because* they did not complete their migration from nose to brain. They did not complete their migration *because* of a mutation in the Kallmann gene. Thus, this Syndrome shows us a clear example of a Gene/Behavior causal relation in humans, but the causal relation goes through no less than six causal steps.

Sex hormone receptors. Here is the most straightforward story of how sex hormone signaling works, not just in the brain but in all hormone-sensitive organs including human breast, uterus, prostate gland, muscles.

Many hormones have the chemical structures of steroids, each steroid having four stable rings of carbons. Steroid hormones are distinguished from each other—estrogens, androgens (like testosterone), and others such as a stress hormone cortisol—from each other by different chemical side chains hanging off the carbon rings like jewels on a necklace. For example, once estrogenic hormones have bound to a protein called Estrogen Receptor-alpha, that protein races to the cell nucleus and, with remarkable speed binds to a special portion of DNA. The simplest version of that portion of DNA, called the Estrogen Response Element has the nucleotide bases AGGTCAnnnTGACCT. Those "n" nucleotide bases can be any one of the basic four, but there have to be three of them in order to allow the Estrogen Response element. Notice that in DNA a "T" pairs together with an "A" and a "C" pairs together with a "G." Now take a look at the Estrogen Response Element and notice that the bases on the left side of the "nnn" match up with the bases on the right side, thus to form a hairpin-like protrusion. To the two sides of that protrusion, two molecules of Estrogen Receptor-alpha bind. Once they have bound, coupled with favorable molecular genetic conditions in that region of the nerve cell nucleus, estrogen-sensitive genes are transcribed (Fig. 3).

How does that affect social behavior? Well, estrogen receptors were discovered in neurons (Pfaff, 1968; Pfaff and Keiner, 1973), the first hormone receptor genes found expressed in brain, and as a result of estrogen binding, female sociosexual behaviors can appear (Pfaff, 2017).

It is roughly similar for males. In males some social behaviors such as sex and aggression (competing for females) depend on these molecular genetic and neurochemical steps: The hormone testosterone has an Androgen Response Element whose nucleotide base sequence is AGAACAnnnTGTTCT, and after two Androgen Receptor proteins

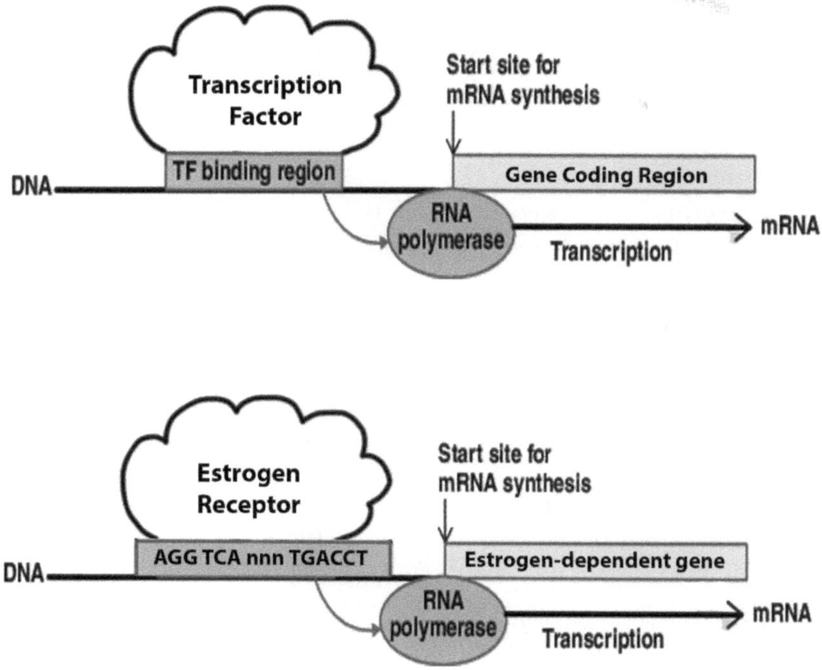

Fig. 3 Cartoon of estrogenic hormone-regulated transcription, as might be important for some social behaviors. When bound by the steroid hormone estradiol, estrogen receptor proteins bind as "dimers" (pairs) to the two sides of a DNA "estrogen response element (ERE)," thus to facilitate transcription. The specific nucleotide base sequence shown comprises the "consensus" ERE sequence, variations upon which are allowed. Androgenic hormones work in an analogous way.

(loaded with testosterone) bind to this hairpin, androgen-dependent behaviors can be facilitated (reviewed, Pfaff and Joels, 2017). One of the complexities I am skipping over is that some androgen dependent behaviors also require chemical conversion of testosterone to estrogen! Another complexity is that different hormone receptors can bind to that same nucleotide base sequence, and the differences among the molecular chemical effects of those different hormone receptors depend on that sequence's flanking regions.

Ultimately, for mammalian sexual behavior which is the key to reproductive success, the evolutionary question boils down to the evolution of socially important gene regulatory networks, but much work along these lines remains to be done (see New questions).

How an individual neurochemical's gene contributes to sociality is a story best told by the neuropeptide oxytocin.

Oxytocin

In the popular press, oxytocin is linked time and again to various forms of human and animal social behaviors, from pair-bonding in tiny wild rodents to memory of social encounters by laboratory mice, to human sexuality and human parenting

The neuropeptide oxytocin itself is a string of nine amino acids twisted into a peculiar shape by a chemical bond composed of two sulfur atoms. The oxytocin gene—the gene that codes for making oxytocin—is expressed in a small crescent of giant neurons in the front part of the hypothalamus (Rhodes et al., 1981). Some of these neurons, for example in the supraoptic nucleus, send axons mainly to the posterior pituitary gland from which oxytocin is secreted into the blood stream for distribution throughout the body. Others of these neurons, for example in the paraventricular nucleus of the hypothalamus, project not only to the posterior pituitary but also widely throughout the brain (Conrad and Pfaff, 1975). These oxytocinergic neuronal axons to the forebrain and to the brainstem offer the route by which oxytocin gene expression is translated into social behavioral modification.

Larry Young, professor at Emory University, has carried forward the most vigorous program of molecular genetic research into exactly how oxytocin promotes prosocial behaviors (reviewed, Walum and Young, 2018). Oxytocin is encoded in a single gene with three distinct exons. The oxytocin unit of the gene is on exon A, and the bulk of the gene codes for a protein called a neurophysin, which is a carrier protein important for oxytocin transport in the neuron. Dietmar Richter's lab in Hamburg, Germany (Mohr et al., 1985) found exactly the same gene structure in the human hypothalamus as had been found in rat brain and cow brain. Following transport, the next step is to analyze the receptor types to which oxytocin can bind (see New questions). It turns out that, just like New York real estate, it is a matter of location, location, location. One species of a tiny mammal called voles has receptor expression in a specific area of the basal forebrain, while the other species does not. In parallel, one of these species is "highly social and monogamous, while others are relatively asocial and nonmonogamous." Subsequent work in Young's lab showed that the molecular and neuroanatomical differences between the two species are causal to the behavioral differences, not merely correlational. Some of the most exotic molecular genetic approaches to these interspecies differences in social behavior have to do with the promoter regions which control transcription of the appropriate receptor genes (reviewed, Hammock et al., 2007).

In these gene regulatory regions there are strings of highly repetitive DNA sequences called "microsatellites," of the sort common throughout mammalian genomes and highly subject to mutation. Elizabeth Hammock, now at Florida State University, and Larry Young did "initial experiments using serial 'nested' deletions through a region of 3500 nucleotide DNA bases surrounding prairie vole gene named avpr1a 5′, with results which indicated that the microsatellite region contained regulatory elements." These elements only modified activity in some cell lines. Further inquiry, utilizing targeted deletion of the microsatellite, revealed that the specific deletion of the microsatellite also modified gene expression in a cell type-dependent manner. Finally, "exchanging the large prairie vole species microsatellite for the small montane species microsatellite, in the surrounding 3.5 kb of prairie vole avpr1a 5′ region, also resulted in altered gene expression" (Hammock and Young, 2005). These microsatellite differences drive gene expression and, with other relevant pharmacological results, are proved to constitute a significant modulator of species differences in social behavior.

Further, what about oxytocin release from the neuron's synapses? Larry Young (2007) discussed a mystifying discovery that a gene named CD38 regulates the release of OT from hypothalamic neurons. This gene codes for a protein which crosses the cell membrane, reasonably enough, but it had previously been known for triggering proliferation and immune responses in blood cells, lymphocytes. Young's discussion exemplifies a classic situation in modern behavior genetics: we wade into the genetic literature with discoveries of behavioral importance, in this case social behaviors, and are flummoxed by the rich and unexpected scholarly history of the genes with which we have to deal.

The social behavioral importance of all of the molecular genetic work above is shown by the paper of Ferguson et al. (2000) who found that "male mice mutant for the oxytocin gene ($Oxt^{-/-}$) failed to develop social memory, whereas wild-type ($Oxt^{+/+}$) mice showed intact social memory" (memory for a previous social partner). Oxytocin as administered rescued social memory in the mutant animals. Ferguson et al concluded that "data indicate that OT is necessary for the normal development of social memory in mice and support the hypothesis that social memory has a neural basis distinct from other forms of memory." Ferguson and Young's results perfectly match conclusions that Elena Choleris (now at Guelph University in Canada) and I reached (Choleris et al., 2003, 2006). A four gene micronet operating in hypothalamus and amygdala—the four genes being oxytocin, oxytocin receptor, and estrogen receptors alpha and beta,—are necessary for normal social recognition and memory.

Exactly *how* does oxytocin gene expression in neurons of the hypothalamus promote social behaviors? Well, because of the complexity of oxytocin gene regulations and because of widespread distribution of oxytocin neuron axons in the brain, I can anticipate that there are multiple answers to this question. I will highlight just two. First, a causal route which uses neurons in the biologically ancient forebrain, has to do with the blocking of stress effects. Second, molecular genetic studies of neurons in the biologically newest and most sophisticated parts of the cerebral cortex have yielded different, complementary answers.

First, in 1991, Margaret McCarthy and I theorized that a role for oxytocin to promote reproductive behaviors is to protect such behaviors from disruption by stress. That turned out to be true. Intranasal infusion of oxytocin appears to enhance the buffering of social support in humans, as indicated by decreased stress response during a socially stressful situation (Heinrichs et al., 2003). And in Tohoku University, Sendai, Japan, the lab of Kenji Nishimori (Yoshida et al., 2009), oxytocin receptor gene expression has been implicated in the reduction of anxiety. These scientists "found oxytocin receptor expression in about one-half of tryptophan hydroxylase gene expressing neurons important for making the emotionally powerful neurotransmitter serotonin. Oxytocin infusion facilitated serotonin release within the median raphe nucleus of the midbrain and reduced anxiety-related behavior. Infusion of a serotonin receptor antagonist blocked the anxiolytic effect of oxytocin, suggesting that oxytocin receptor activation in serotonergic neurons mediates the anxiolytic effects of oxytocin." So, oxytocin's stress reduction can be tied, in part, to an anti-anxiety neurotransmitter, serotonin.

Second, mechanisms for oxytocin's behavioral effects include the prefrontal cortex. How does oxytocin get to the prefrontal cortex? Oxytocin does not cross the blood brain barrier and most neuroanatomical studies do not report cortical projections. However, Robert Froemke's lab at New York University (Mitre et al., 2017) employed an unusual technique and did get evidence of a cortical projection. They injected a special viral vector in which a colored marker was expressed under the control of the mouse oxytocin gene promoter, "into the sites of oxytocin gene expression in the hypothalamus. As a result, fluorescent fibers were detected as immunostained with antibodies to yellow fluorescent proteins, in the frontal and orbital frontal cortex". They focused on social cognition manifest in the mother animal's cortical response to the alarm calls from her pups. Surprisingly, oxytocin effects depended on what side of the cortex they looked at. "Retrieval behaviour required the left but not right auditory cortex, was accelerated by oxytocin in the left auditory cortex, and oxytocin receptors were preferentially expressed in the left auditory cortex. Neural responses to pup calls were lateralized, with co-tuned and temporally precise excitatory and inhibitory responses in the left cortex of maternal but not pup-naive adults" (Marlin et al., 2015). Structural and electrical studies supported this Marlin report. "Electron microscopic analysis of the cerebral cortex revealed that oxytocin receptors were mainly expressed at synapses, as well as on axons and glial processes. Functionally, oxytocin transiently reduced synaptic inhibition in multiple brain regions and enabled long-term synaptic plasticity in the auditory cortex" (Mitre et al., 2016).

Other experiments demonstrating the importance of the frontal cortex for oxytocin effects used different molecular genetic techniques. Nakajima et al. (2014) discovered a population of inhibitory neurons in the prefrontal cortex. Genetically silencing those neurons "resulted in loss of social interest by the female in male mice specifically during the sexually receptive phase of the estrous cycle." These authors concluded that oxytocin receptor expressing neurons in the medial prefrontal cortex are required for normal female sociosexual behavior during estrus.

Inspired by these molecular genetic studies and by other parts of the literature, many scientists have tried to use intranasal delivery of oxytocin to improve social behavior in nonhuman primates and in human subjects. It turns out that, while exciting, these studies need to be reviewed with an adequate degree of rigor. For example, some experiments with monkeys claim that after intranasal oxytocin, cerebrospinal fluid levels of oxytocin are elevated (Dal Monte et al., 2014; Lee et al., 2018), but the lab of the eminent behavioral biologist Frans deWaal (Proctor et al., 2016) failed to see behavioral effects of intranasal oxytocin in chimpanzees, our closest living relatives. Oxytocin expert at the Edinburgh University Gareth Leng, with Mike Ludwig (2016) looked at the claims about oxytocin passage from the nose using quantitative thinking and these authors worry about interpretative mistakes in the behavioral studies. For example, Leng and Ludwig state that, following intranasal delivery, oxytocin concentrations were high in small regions known to lack a blood-brain barrier, but deeper and broader penetration into the brain was absent. So the interpretation of intranasal oxytocin studies remains up in the air.

Above, I have been discussing genes whose products facilitate social interactions, but it is time to reflect on genes associated with miserable social experiences.

Social problems

Here I will talk about two examples of the intersections between modern genetics and problems with social behaviors. The first example starts from the individual himself, autism. The second example reflects how the environment can weigh upon an individual's opportunity for social interactions.

Autism. Adding to what I covered in Chapter 9, I warn you that nothing about genetics and autism is simple. Almost 800 gene mutations have been associated with autism. Let's look at just one of these.

Some of the most convincing examples of genes whose mutations pose risks for autism are genes whose products foster normal synaptic communication. An early-studied example of these is a member of the SHANK gene family of synaptic proteins, which supply "scaffolds" in the postsynaptic density for holding postsynaptic ion channels and neurotransmitters in place. Following work by Moessner et al. (2007) and Durand et al. (2007) establishing a connection between one SHANK protein, SHANK3 and autism, the team of Joseph Buxbaum, a geneticist at Mount Sinai Medical School (Bozdagi et al., 2010; Betancur and Buxbaum, 2013) analyzed SHANK3's contributions to social behavior, in a neurobiological context.

Recently a team at the Chinese Academy of Sciences laboratory in Shenzhen (Zhou et al., 2019) reported interesting results monkey brains in which genetic engineering had produced specific types of mutations in SHANK3. These monkeys with reduced expression of SHANK3 showed sleep disturbances, motor deficits and increased repetitive behaviors, and social impairments reminiscent in some respects with autism. In addition to genes coding for synaptic proteins, genes whose products are involved in transcription regulation have mutations strongly associated with autism (reviewed, De Rubeis et al., 2014).

Loneliness. Loss of a social life often is imposed on people, and they feel lonely. In Chapter 10, I talked about the work of the Cacioppo team at the University of Chicago, notably how they reported physiological effects of loneliness which they interpreted as bad effects. Certainly, the down-regulation, the reduction of expression of interferon anti-viral genes can be counted as bad. So, with a tremendous amount of humility as we face an interferon gene literature of astonishing complexity and multiplicity—a huge field of scientific and medical work—let's try to see what the Cacioppo team's findings mean, genetically and medically. To do so I must deal with a complex subject: interferon anti-viral genes.

As summarized by Hardy (2004), some human genes, including interferon genes, are located in clusters on given chromosomes. For example, for Type I interferons, their genes are clustered on a specific part of human chromosome 9, using up a (relatively small) space of 400,000 nucleotide bases. Most genes typically have segments coding for messenger RNAs and proteins (segments called "exons") separated by non-coding "introns." But Hardy et al. discuss interferon genes which have no introns. The consequences of this are unclear.

Interferon genes produce barrel-shaped proteins whose evolutionary histories are complex. For example, Hardy et al. reveal an evolutionary tree for one interferon which has no less than 48 forebearers. The genomic regulations over interferon gene expression are complex and variegated, but we know that the end goal is to permit defense against pathogens, mostly viruses. In response to such pathogens interferons are released from cells—for example blood cells and fibroblasts (connective tissue). One consequence of special interest in neuroscience is that interferon type-alpha affects nerve cells in a part of the anterior hypothalamus, the preoptic area, nerve cells that control body temperature. The interferon effect there allows body temperature to go up—fever—to temperatures which kill pathogens.

Thus, down-regulation, reduction of gene expression for interferons, or interference with their actions certainly is unfortunate, presumably rendering people suffering longterm from loneliness more susceptible to viral infections.

Parent-of-origin genomic imprinting in the brain

Virtually all genes have two copies, one from the father (in the sperm) and the other copy from the mother (in the egg). There are some genes whose expression greatly favors the copy from the father, and others from mother. According to a review by Tucci, Ferguson-Smith et al. (2019), about 67 favor the father's copy, and 21 the mother's.

How does this happen? Parent-of-origin genomic imprinting comes from chemicals added on to DNA or to the proteins which cover up DNA in the cell nucleus. These are "epigenetic" mechanisms which affect gene expression without modification of the order of nucleotide bases (C, G, A, T) on the gene. The dominant mechanism is "DNA methylation—the addition of a methyl (CH_3) group to the C's of DNA - - it is a heritable epigenetic mark that is associated with transcriptional repression when occurring on regulatory sequences, such as promoters. The parent-of-origin-specific acquisition of DNA methylation marks at imprinting control regions (ICRs) in the germline is essential for the mono-allelic *(father's or mother's copy)* expression of imprinted genes in embryos and are retained as the memory of parental origin after fertilization" (Tucci, Ferguson-Smith et al., 2019) (italics mine).

Social behaviors. Back in a time when none of us were thinking about parent-of-origin controls over gene expression, E. Barry Keverne, professor at the University of Cambridge and an expert in social behaviors by animals, conceived the importance of this subject. To begin with, charting brain development, he (Keverne et al., 1996) reported that father-originated "cells (with duplicated paternal genomes) contributed substantially to the hypothalamic structures and not the cortex. By contrast, mother-originated cells (with duplicated maternal genomes) contributed substantially to the cortex, striatum and hippocampus but not to the hypothalamic structures." Keverne also foresaw roles for genetic imprinting in certain diseases such as Angelman syndrome (intellectual disabilities), bipolar depression and schizophrenia (Keverne, 1997). Isles et al. (2006) focused on social bonding and social behaviors. According to these authors, imprinted genes mediate these behaviors, including mother-offspring interactions and certain adult social interactions.

This field of parent-of-origin genomic imprinting is relatively new in behavioral neuroscience and is developing slowly because, in the lab, experiments are hard and, in society, opportunities for systematic studies are sparse. Direct experimental proof of causal links to behavior is still lacking. But I predict as genomic methodologies improve and as the public becomes used to be examined for gene/behavior relations, the importance of genomic imprinting will be realized.

In this rapid, "36,000 ft" view of how certain genes have fitted into the subjects of this book, I now come to a relatively new subject: modification of gene expression while leaving the four nucleotide bases of DNA itself in exactly the same coding sequence.

Epigenetics

"Epigenetics" refers to all molecular processes which can affect efficiency of gene transcription without altering the nucleotide (A, T, G, C) base sequence at all (Allis and Jenuwein, 2016; Jenuwein and Allis, 2001). In reality, demonstrations of epigenetic power have been around more than 50 years (Allfrey et al., 1964). And we should not be surprised. It is worth it to remember that, before DNA was discovered as the genetic substance by Oswald Avery (it took him 17 years, from 1927 to 1944!), geneticists figured that a simple, 4-base linear code is simply "too stupid" to constitute the genetic substance. They were right. The ability to modulate transcription efficiency without wrecking the DNA itself exponentially increases the organism's genomics' abilities to carry high information content. Consider epigenetics, from the greek "epi-" (upon, beyond, after) genetics.

Here's the key to epigenetics. When DNA is inactive, tied up, not transcribing it is usually wound up by proteins in little bundles called nucleosomes. But the tails of some of those proteins—called histones, which are basic to bind to the acidic DNA—stick out of the nucleosomes and are subject to chemical modification. Here is how the modification works. There are some enzymes which my colleague David Allis calls "writers" which apply the chemical modification to the histone protein tail. Then there are enzymes, "readers," which recognize and use the chemical modification in a manner that depends on what exactly that modification is and exactly where it is on the protein tail. This is very specific. Every atom counts. Finally, there are enzymes which are "erasers" which remove the modification.

If the chemical modification is an acetyl group (CH_3CO), almost always that histone tail modification favors transcription of the nearby gene. If the chemical modification is a methyl group (CH_3), that tail modification usually blocks transcription, except for some specific histone tail locations.

Epigenetics in social behavior. Most important: because of these epigenetic mechanisms, for social behavior, organisms including humans and lab animals can have their genetic outputs *modified by social experience.*

A review by Bludau et al. (2019) from Inga Neumann's department at the University of Regensburg focuses on the role of the epigenetic network in regulating the enduring effects of social experiences during early-life, adolescence, and adulthood. For example, they refer to fetal alcohol exposure "leading to global hypermethylation, reduced levels of certain epigenetic additions to histone proteins, decreasing the levels of β-endorphin in the hypothalamus" β-Endorphin is a key regulator of social affect and attachment, evidenced by its prosocial effects after social isolation in rats.

In my lab, Schaafsma et al. (2017) used inflammatory stress during the first trimester to test a "three-hit" theory (genetic load × environmental stress × sex) of autism. We found that maternal immune activation caused male-specific deficits in certain social responses in the contactin-associated protein-like 2 (Cntnap2 gene mutation) mouse model of autism. The three "hits" had cumulative effects on ultrasonic vocalizations at postnatal day 3. Hits synergistically affected social recognition in adulthood: only mice exposed to all three hits showed deficits in this aspect of social behavior. In brains of the same mice we found a significant three-way interaction on corticotropin-releasing hormone receptor-1 (Crhr1) gene expression, in the left hippocampus specifically, which co-occurred with epigenetic alterations in histone H3 N-terminal lysine 4 trimethylation (H3K4me3) over the Crhr1 promoter. This finding is especially interesting because corticotropin-releasing hormone is a kingpin in the brain's response to stress. So this experiment drew in a specific epigenetic reaction in a specific part of the brain as it might contribute to social deficits in autism.

What about the elemental social behaviors of sex and maternal care? Khatuna Gagnidze, in my lab (2013) investigated epigenetic chemistry as it could contribute to the estrogenic facilitation of female reproductive behavior. This behavior depends on estrogen-induced expression of progesterone receptor (PgR) within a defined cell group in the ventrolateral portion of the ventromedial nucleus of the hypothalamus (VMH). Estradiol administration induced rapid and transient global histone modifications in the VMH of ovariectomised female mice. For example, histone H3 N-terminus phosphorylation, acetylation and methylation (H3K4me3) exhibited distinct temporal patterns facilitative to the induction of transcription. Moreover, examination of histone modifications associated with the promoter of another estrogen receptor-target gene, oxytocin receptor (Oxtr), revealed gene- and brain-region specific effects of estrogenic treatment. By no means do we know the entire "histone code" for estrogenic effects on sex behavior, but these results provided proof of principle.

With respect to maternal care, Frances Champagne (2012) has investigated epigenetic endpoints as they relate to different levels of adequacy of maternal care. For example, according to her review, poor maternal care "alters DNA methylation of multiple genes involved in the regulation of brain development, function, and behavior. For example, exaggerated stress responsivity in the offspring of low care dams is suggested to be due to DNA methyltransferase 1 (DNMT1)-mediated promoter hypermethylation of genes such as the glucocorticoid receptor (GR) gene (*Nr3c1*) within the hippocampus. This reduces GR expression and impairs its negative feedback on the hypothalamus–pituitary–adrenal (HPA) axis, which consequently increases stress responsiveness. Furthermore, such heightened stress reactivity severely diminishes the quality of maternal care displayed by the adult female offspring of low-LG dams."Thus, an epigenetically mediated positive feedback whose implications carry forward into the next generation.

In summary, during the last 20 years or so, epigenetic mechanisms have leapt into the scientific spotlight (Grunstein and Allis, 2018), and our understanding of its role in human social phenomena will grow exponentially.

New questions

1. Yes, GnRH is the "master molecule" promoting many steps of reproduction in both sexes, but another molecule called "kisspeptin" regulates GnRH neurons. Exactly how kisspeptin does that constitutes a currently active area of research.
2. Oxytocin is the famous prosocial molecule, but its relation to a very similar nine amino acid neuropeptide can be a bit confusing. The two genes, oxytocin and vasopressin, are on the same chromosome separated by DNA which includes a LINE element, a Long Interspersed Nuclear Element, an exotic piece of a transposon, a "jumping gene." What this means for oxytocin/vasopressin neurophysiology remains fascinating but obscure. Further, although vasopressin, in some studies, fosters aggressive behaviors, oxytocin can act through a vasopressin receptor. Along those lines, in my lab Andre Ragnauth (2004), using electrical recordings from individual neurons, showed that vasopressin stimulates ventromedial hypothalamic neurons via oxytocin receptors. How did this come about?
3. Larry Young has envisioned how oxytocin systems might be related to dopamine-responsive systems in the forebrain connected with behavioral reward. A question comes up as to how such connections are related to the Approach/Avoidance mechanisms Helen Barbas and I described in Chapter 9.
4. Forkhead gene FoxP2 clearly has something to do with human speech. But forkhead genes are really complicated. Is their function on speech development really just a special case of motor learning?
5. Chemically coded networks: How can they change our thinking?
   With data from lobsters, nematodes, fruit flies, and the retina, molecular neurobiologist Cori Bargmann (2012) illustrated how the effective performance of neural circuits could be "actively shaped by neuromodulators that modify neuronal dynamics, excitability, and synaptic function".

Such neuromodulators could include serotonin, dopamine, tyramine, and octopamine, as well as small pieces of proteins, neuropeptides. Latent circuits, sometimes activated, sometimes not. In fact, my lab's results had already attested to the cogency of Bargmann's idea. We found that groups of neurons which express the Estrogen Receptor-alpha gene strongly tend to project axons to other neuronal groups which express the Estrogen Receptor-alpha gene (Pfaff and Keiner, 1973; Cottingham and Pfaff, 1986). Thus, a latent hormone-sensitive circuit in which hormone effects on behavioral performance could be increased additively or multiplicatively by the administration of estrogenic hormones.

What about gene networks? Baran et al. (2017) have done a service to the field by applying network thinking, which neuroscientists do naturally as just illustrated, to the evolution of social behavior. They énvision gene regulatory networks and apply that thinking to the vertebrate social behavior network. And, in their words "The big picture that we are striving for will show how regulatory relations among genes/gene products play into traditionally defined neuronal networks for social behaviors (as detailed in previous chapters); and, in turn, how these genetic and neuronal realities play into the dynamics of biologically adaptive social behaviors." During coming years, can we neuroscientists and geneticists, working hand in hand with mathematicians, begin to fill in the necessary details? In fact, the laboratory of Matt State (Willsey, 2013) has already used gene co-expression networks to zoom in on human frontal cortical neurons as important for the genesis of autism. Starting with nine high-confidence autism-related gene mutations they came up with human midfetal layer 5/6 cortical projection neurons as key to the development of the disease. So, in coming years, can neuroscientists and geneticists do the high wire act of thinking quantitatively about gene networks and neural networks in the same intellectual framework in order to deepen our understanding of social behavior mechanisms?

6. Consider gene therapy. We are all very familiar with the agonizing situation in which a single gene mutation can cause a tragic rare disease. What to do about it? A few well-known medical geneticists have advocated simply changing the disease-causing DNA: substitute DNA, remove DNA, add DNA. But can we really estimate the probability of unintended consequences? After all, the cell nucleus is a very crowded and complicated place. For example, RNA biology provides numerous examples of sophisticated biochemical processing (examples in Darnell, 2013; Jain et al., 2020), which could be altered by genetic engineering, by mistake.

The simplest criticisms simply ask whether the geneticist can do what (s) he intends and nothing but. Will the desired genetic change actually be achieved? And what about mistakes called "off-target effects." They may be unsafe. As well, consider unpredictable genome-wide effects "where changes to one gene alter the expression of many genes in unpredictable ways" (Baylis, 2019, p. 23). And after an attempt at genetic manipulation, there may be a mixture of corrected and uncorrected cells. A leading stem cell biologist at MIT, Rudolf Jaenisch, is quoted as doubting "that researchers will ever be able to rule out the possibility of mosaicism (an uncontrolled mixture of affected and non-affected cells) in an embryo." Doctors will not even know whether enough cells are corrected to treat the disease.

Dalhousie University professor Francoise Baylis (2019) quotes master geneticist Eric Lander (p. 940): "We've only just skimmed the three-billion-yea-old genetic text and already they're saying 'Hey! I think I can improve on this!'" And (p. 118) the director of the NIH Francis Collins: "Evolution has been working toward optimizing the human genome for 3.85 billion years. Do we really think that some small group of human genome tinkerers could do better without all sorts of unintended consequences?"

To the point, will this kind of genetic thinking ever be applied to behavior, even social behavior?

References

Allfrey, V.G., Faulkner, R., Mirsky, A.E., 1964. Acetylation and methylation of histones and their possible role in the regulation of RNA synthesis. Proc. Natl. Acad. Sci. U. S. A. 51, 786–794.

Allis, C.D., Jenuwein, T., 2016. The molecular hallmarks of epigenetic control. Nat. Rev. Genet. 17 (8), 487–500.

Baran, N.M., McGrath, P.T., Streelman, J.T., 2017. Applying gene regulatory network logic to the evolution of social behavior. Proc. Natl. Acad. Sci. 114 (23), 5886–5893.

Bargmann, C.I., 2012. Beyond the connectome: how neuromodulators shape neural circuits. BioEssays 34 (6), 458–465.

Baylis, F., 2019. Altered Inheritance. Harvard University Press, Cambridge.

Belsham, D.D., Lovejoy, D.A., 2005. Gonadotropin-releasing hormone: gene evolution, expression, and regulation. Vitam. Horm. 71, 59–94.

Betancur, C., Buxbaum, J.D., 2013. SHANK3 Haploinsufficiency: a "common" but underdiagnosed highly penetrant monogenic cause of autism spectrum disorders. Mol. Autism 4 (1), 17–31.

Bludau, A., et al., 2019. Epigenetic regulation of the social brain. Trends Neurosci. 42 (7), 471–484.

Bozdagi, O., et al., 2010. Haploinsufficiency of the autism-associated Shank3 gene leads to deficits in synaptic function, social interaction, and social communication. Mol. Autism 1, 15. PMID: 21167025.

Bray, S.J., 2016. Notch signalling in context. Nat. Rev. Mol. Cell Biol. 17, 722–735.
Champagne, F.A., 2012. Interplay between social experiences and the genome: epigenetic consequences for behavior. Adv. Genet. 77, 33–57.
Choleris, E., et al., 2003. An estrogen-dependent four-gene micronet regulating social recognition: a study with oxytocin and estrogen receptor-alpha and -beta knockout mice. Proc. Natl. Acad. Sci. U. S. A. 100 (10), 6192–6197.
Choleris, E., et al., 2006. Involvement of estrogen receptor alpha, beta and oxytocin in social discrimination: a detailed behavioral analysis with knockout female mice. Genes Brain Behav. 5 (7), 528–539.
Conrad, L.A., Pfaff, D.W., 1975. Axonal projections of medial preoptic and anterior hypothalamic neurons. Science 190, 1112–1114.
Cottingham, S.L., Pfaff, D.W., 1986. Interconnectedness of steroid hormonebinding neurons: existence and implications. In: Ganten, D., Pfaff, D.W. (Eds.), Current Topics of Neuroendocrinology. 7, pp. 223–250.
Dal Monte, O., et al., 2014. CSF and blood oxytocin concentration changes following intranasal delivery in macaque. PLoS One 9 (8), e103677.
Darnell, R.B., 2013. RNA protein interaction in neurons. Annu. Rev. Neurosci. 36, 243–270.
De Rubeis, S., et al., 2014. Synaptic, transcriptional and chromatin genes disrupted in autism. Nature 515 (7526), 209–215.
Durand, C.M., et al., 2007. Mutations in the gene encoding the synaptic scaffolding protein SHANK3 are associated with autism spectrum disorders. Nat. Genet. 39 (1), 25–27.
Ferguson, J.N., et al., 2000. Social amnesia in mice lacking the oxytocin gene. Nat. Genet. 25 (3), 284–288.
Fisher, S.E., Scharff, C., 2009. FOXP2 as a molecular window into speech and language. Trends Genet. 25 (4), 166–177.
Gagnidze, K., et al., 2013. Early histone modifications in the ventromedial hypothalamus and preoptic area following oestradiol administration. Neuroendocrinology 25 (10), 939–955.
Grunstein, M., Allis, C.D., 2018. Genetics, biochemistry, and "simple" organisms converge to unlock secrets in histone biology. JAMA 320 (12), 1233–1234.
Hammock, E.A.D., Young, L.J., 2005. Microsatellite instability generates diversity in brain and sociobehavior. Science 308 (5728), 1630–1634.
Hammock, E.A.D., et al., 2007. Gene regulation as a modulator of social preference in voles. Adv. Genet. 59, 107–119.
Hardy, M.P., 2004. Characterization of the type I interferon locus and identification of novel genes. Genomics 84, 331–345.
Heinrichs, M., Baumgartner, T., Kirschbaum, C., Ehlert, U., 2003. Social support and oxytocin interact to suppress cortisol and subjective responses to psychosocial stress. Biol. Psychiatry 54, 1389–1398.
Hevner, R.F., 2019. Intermediate progenitors and Tbr2 in cortical development. J. Anat. 235 (3), 616–625.
Isles, A.R., Davies, W., Wilkinson, L.S., 2006. Genomic imprinting and the social brain. Philos. Trans. R. Soc. Lond. Ser. B Biol. Sci. 361 (1476), 2229–2237.
Jain, N., et al., 2020. Transcription polymerase catalyzed emergence of novel RNA replicons. Science 368, 153–158.
Jenuwein, T., Allis, C.D., 2001. Translating the histone code. Science 293 (5532), 1074–1080.
Keverne, E.B., 1997. Genomic imprinting in the brain. Curr. Opin. Neurobiol. 7 (4), 463–468.
Keverne, E.B., et al., 1996. Genomic imprinting and the differential roles of parental genomes in brain development. Dev. Brain Res. 92 (1), 91–100.
Kim, Y.J., et al., 2018. WDR11-mediated hedgehog signalling defects underlie a new ciliopathy related to Kallmann syndrome. EMBO Rep. 19 (2), 269–289.

Lee, M.R., et al., 2018. Oxytocin by intranasal and intravenous routes reaches the cerebrospinal fluid in rhesus macaques: determination using a novel oxytocin assay. Mol. Psychiatry 23 (1), 115–122.
Leng, G., Ludwig, M., 2016. Intranasal oxytocin: myths and delusions. Biol. Psychiatry 79, 243–250.
Marlin, B.J., et al., 2015. Oxytocin enables maternal behaviour by balancing cortical inhibition. Nature 520 (7548), 499–504.
Mitre, M., et al., 2016. A distributed network for social cognition enriched for oxytocin receptors. J. Neurosci. 36 (8), 2517–2535.
Mitre, M., et al., 2017. Oxytocin modulation of neural circuits. Curr. Top. Behav. Neurosci. 35, 31–53.
Moessner, R., et al., 2007. Contribution of SHANK3 mutations to autism Spectrum disorder. Am. J. Hum. Genet. 81 (6), 1289–1297.
Mohr, E., et al., 1985. Expression of the vasopressin and oxytocin genes in human hypothalami. FEBS Lett. 193 (1), 12–16.
Nakajima, M., Görlich, A., Heintz, N., 2014. Oxytocin modulates female sociosexual behavior through a specific class of prefrontal cortical interneurons. Cell 159 (2), 295–305.
Parhar, I., et al., 1995. Embryonic development of gonadotropin-releasing hormone neurons in the sockeye salmon. J. Comp. Neurol. 362 (2), 256–270.
Parhar, I., et al., 1996. Gonadotropin-releasing hormone gene expression in teleosts. Mol. Brain Res. 41 (1–2), 216–227.
Pfaff, D.W., 1968. Autoradiographic localization of radioactivity in rat brain after injection of tritiated sex hormones. Science 161, 1355–1356.
Pfaff, D.W., 1973. Luteinizing hormone releasing factor (LRF) potentiates lordosis behavior in hypophysectomized ovariectomized female rats. Science 182, 1148–1149.
Pfaff, D.W., 2017. How the Vertebrate Brain Regulates Behavior: Direct from the Lab. Harvard University Press, Cambridge, MA.
Pfaff, D.W., Joels, M. (Eds.), 2017. Hormones, Brain and Behavior, third ed. Elsevier, Cambridge.
Pfaff, D.W., Keiner, M., 1973. Atlas of estradiol-concentrating cells in the central nervous system of the female rat. J. Comp. Neurol. 151, 121–158.
Proctor, D., et al., 2016. Intranasal oxytocin failed to affect chimpanzee (pan troglodytes) social behavior. Anim. Behav. Cogn. 3 (3), 150–158.
Ragnauth, A., et al., 2004. Vasopressin stimulates ventromedial hypothalamic neurons via oxytocin receptors in oxytocin gene knockout male and female mice. Neuroendocrinology 80, 92–99.
Rhodes, C.H., et al., 1981. Immunohistochemical analysis of magnocellular elements in rat hypothalamus: distribution and numbers of cells containing neurophysin, oxytocin, and vasopressin. J. Comp. Neurol. 198, 45–64.
Schwanzel-Fukuda, M., Pfaff, D.W., 1989. Origin of luteinizing hormone-releasing hormone neurons. Nature 338, 161–164.
Schwanzel-Fukuda, M., Bick, D., Pfaff, D.W., 1989. Luteinizing hormone-releasing hormone (LHRH)-expressing cells do not migrate normally in an inherited hypogonadal (Kallmann) syndrome. Mol. Brain Res. 6, 311–326.
Schaafsma, S.M., et al., 2017. Sex-specific gene–environment interactions underlying ASD-like behaviors. PNAS 114 (6), 1383–1388.
Schwanzel-Fukuda, M., et al., 1996. Migration of LHRH neurons in early human embryos: association with neural cell adhesion molecules. J. Comp. Neurol. 366, 547–557.
Stamou, M.I., Georgopoulos, N.A., 2018. Kallmann syndrome: phenotype and genotype of hypogonadotropic hypogonadism. Metabolism 86, 124–134.
Tucci, V., et al., 2019. Genomic imprinting and physiological processes in mammals. Cell 176 (5), 952–965.

Walum, H., Young, L.J., 2018. The neural mechanisms and circuitry of the pair bond. Nat. Rev. Neurosci. 19 (11), 643–654.
Willsey, A.J., 2013. Coexpression networks implicate human midfetal deep cortical projection neurons in the pathogenesis of autism. Cell 155 (5), 997–1007.
Yoshida, M., et al., 2009. Evidence that oxytocin exerts anxiolytic effects via oxytocin receptor expressed in serotonergic neurons in mice. J. Neurosci. 29 (7), 2259–2271.
Young, L.J., 2007. Regulating the social brain: a new role for CD38. Neuron 54 (3), 353–356.
Zhou, Y., et al., 2019. Atypical behavior and connectivity in SHANK3-mutant macaques. Nature 570 (7761), 326–331.

CHAPTER 12

Wrap-up and outlook

In "Origins," I have sought to understand the first spark of human attraction towards another human, i.e., the instinct towards social contact. Data relevant to the time scales underlying this impulse are measured in milliseconds, as in neurophysiology; in months, as in babies' development; and in the eons over which *Homo sapiens* evolved, measured to the extent that it can be by paleoanthropology. Accordingly, my book scales across timeframes appropriate to its subjects.

While I was writing, I was "inside a kaleidoscope" of shifting temporal and topical planes. Outside, the coronavirus was upsetting the balance of everything, challenging us to retain all that I was claiming to be natural and normal. Amidst the daily gut-wrenching churn, I wrote about prosocial mechanisms that humans have naturally, and about how—to be social in any sense—we have co-evolved with the technology we fell back on to help us through the crisis.

Since I have lived in New York for over 50 years, and lectured in large cities in 29 countries, I have observed how people in dense agglomerations find ways to get along. This book examines the scientific bases for the individual's social approaches which—as I demonstrate—are astonishingly easy. Our brains are predisposed towards the prosocial.

To explain the origins of human sociability I offer three causal routes into human behavior which are not alternative to or independent of each other but, rather, intertwined. These three scholarly routes are close to Nobel-laureate Nikolaas Tinbergen's requirements for understanding behavior: evolution, development and mechanisms. Because I am a neuroscientist and because of the current excitement concerning neural nets and neural biology, my discussion emphasizes brain mechanisms. But, in fact all three prongs of my theory are equal in terms of causality.

Causal roots intertwined. The first chapter of this book reviewed data from specialists in non-human primate and Neanderthal social behavior. These paleoanthropologists chart the transition from hunter/gatherer societies to early settled communities. They describe the process of human and animal

Origins of Human Socialization
https://doi.org/10.1016/B978-0-323-85861-8.00004-9

domestication. In tandem, genes involved in the growth of brain size were operant, and I discussed these in Chapter 11.

In Chapter 2, I discussed research concerning the development of social eye gaze, parental recognition, and shared attention, along with perturbations in these developments leading to autism.

My approach (evident in these first two chapters) is premised on the idea that during the lives of non-human primates and prehumans, what evolved in fact were the developmental programs which produce the brain mechanisms that govern human sociality. Evolution, development and brain mechanisms are the key, coordinate factors. Further, from the pre-term baby through early childhood, these brain mechanisms develop by means of emerging temporal patterns of gene expression and consequent neuroanatomical connections. The three roots of our social instincts—evolution, babies' development, brain mechanisms—are intertwined and, viewed as such, form a more comprehensive approach to the causation of our prosocial propensity than any single one of its constituent approaches.

Brain mechanisms for the initiation of social behaviors. So where does the spark of social interest begin? As far as brain mechanisms are concerned, the story begins with the social behaviors of fish. Neuroscientists have charted the "social brain" of fish, which persists throughout vertebrate evolution; fish in groups can be so sensitive to each other that serious scientists talk about "collective cognition." At this moment, molecular and genetically inclined neurobiologists are figuring out exactly how the six components of the social brains of fish, covered in Chapter 3, actually work together to produce the biologically adaptive social behaviors fish show, even including "collective cognition."

Again, following the scientific strategy of starting with the simplest problems, fantastic progress has been made analyzing mechanisms for simple social behaviors: sex, parenting, and aggression (Chapter 4). The problem of working out mechanisms for laboratory-animal sex behaviors has been solved. Likewise, we can observe parenting behaviors with considerable understanding of mechanisms. Testosterone-fueled aggression, a significant social problem, can now be approached with the tools of molecular genetics. Brain mechanisms underlying these behaviors, discovered in laboratory animals, remain in place during evolution towards the human brain, and now continue to operate underneath multiple, complex layers of cultural influence.

I addressed the supporting roles of sense perception in regulating we move towards each other. Take vision (Chapter 5). In primate brains, certain nerve cells show electrical activity only when the eyes are shown a face. In the prefrontal cortex, groups of neurons are active selectively and specifically when the subject's eyes perceive social interactions. So-called "top down" influences on visual system signaling allow us to understand how cognitive functions such as attention and motivation can influence what we notice.

With regard to audition (Chapter 6), active manipulation of auditory signals takes place in the ears and in the cerebral cortex. Neuroscientists now argue that the genetic support of neural mechanisms discovered for producing birdsong play a part in the support of vocal learning. In humans, this translates (by analogy) into speech. Almost certainly, the difference between humans' ability to understand, learn and produce speech (compared to other animals) is not a dichotomous all-or-nothing shift, but a matter of degree both in terms of gene expression and of details of neural circuitry.

Ever since the work of psychoanalysts like John Bowlby and Nancy Chodorow (Chapter 7), the scientific analysis of human social behavior has grown from a cottage industry dominated by clinicians to become a major part of modern neuroscience. *Origins of Human Socialization* examines these developments, as in the universal human inclination towards empathy. As Chapter 8 shows, altruistic behaviors can be explained in terms of well-understood brain mechanisms. At the other extreme, I discuss the social aloofness, as in autism (Chapter 9). The heterogeneity among autistic children can be laid not only to the wide variety of gene mutations involved but also the different developmental times at which those genes are supposed to perform their normal functions. Chapter 10 examines the harmful physiological consequences of loneliness, and the early steps scientists have taken towards learning the neuroendocrine and molecular genetic steps involved.

My goal throughout this book has been to demystify. I have sought to show that there are neurophysiological bases for our social behaviors that do not depend, in the first instance, on ethical and religious teachings, even though external conditions can, of course, affect how individuals behave in a given setting. I also hope that, wherever there are external or internal conditions which cause individuals to suffer, there may be reason to believe that modern neuroscience will be able to help untangle the detrimental effects and, accordingly, help to find cures.

Looking forward, I wonder if the co-evolution of humans and machines may be accelerated by the coronavirus Covid-19; and if, far from thinking of evolution as something that "happened," if the evolution of our social instinct is as current as our next log-in to Skype, Facebook, or dozens of other platforms transforming the way that we connect.

Our social future: Room for optimism?

Some public intellectuals appear confident about the human brain's capacity to handle current social challenges. In *The Better Angels of our Nature* (2011), Harvard psychologist Steven Pinker observes that various forms of anti-social behavior (e.g., hate crimes) and criminal violence and warfare have decreased over time. Political scientist Rachel Kleinfeld (2018) agrees. Her chart of homicides per 100,000 people in Italy, Germany, Switzerland Scandinavia Netherlands, Belgium and England shows steady declines from the year 1300 to 2010. Battlefield deaths, while not zero, certainly, are now relatively low worldwide (pp. 8–9). The list goes on and on. Richard Wrangham (2019) writes of our peaceable ability to look each other in the eye, control our tempers and avoid aggression towards unfamiliar people.

Economist Julian Lincoln Simon, who wrote, *It's Getting Better All the Time* suggested that with remarkable technological advances, humans would be able to feed, clothe, and supply energy to an ever-growing population for as long as necessary. British biologist and journalist, Matthew Ridley, argues that our economic activities would prove to be an unquenchable source of human prosperity. In *The Rational Optimist* (2010), he lampoons pessimists, and touts improvements in our standard of living.

In *Abundance: The Future Is Better Than You Think* (2012) Peter Diamandis and Steven Kotler speculate that from a biological perspective, it might be most important for humans to "prioritize bad news" in order to survive, even though the real news is actually very good. Philosopher John Rawls (1993/2005) embraces the possibility of "lasting improvement." Their various approaches can be distilled into the prediction that, as a species, we will survive and even prosper.

But I see flaws in their collective outlook, which assumes a steady environmental state, clearly erroneous in the face of accelerating climate change, and fails to take account of a technological environment that has literally become an extension of how we respond to each other and the world. Or, perhaps, have we become an extension of technology? Is there an emerging mismatch between how our brains were designed and our current social technologies? What are we evolving *towards*, and will our immediate needs, especially during an age of coronavirus, outstrip our ultimate survival potential?

Other reasons, for worry

As I hinted briefly in the book, technology is making all of us immediately present to all of us. We are one planetary society, somewhat artificially separated by shaky national boundaries. This is true both in cyberspace, where everybody can communicate with everybody, and on the ground, where a pandemic like coronavirus can spread throughout the world overnight.

Oxford evolutionary biologist Robin Dunbar worries that social brain mechanisms optimal for small societies are badly stretched by huge ones. "My research is concerned with trying to understand the behavioural, cognitive and neuroendocrinological mechanisms that underpin social bonding in primates (in general) and humans (in particular). Understanding these mechanisms, and the functions that relationships serve, will give us insights how humans have managed to create large scale societies using a form of psychological that is evolutionarily adapted to very small-scale societies, and why these mechanisms are less than perfect in the modern world. This has implications for the design of social networking sites as well as mobile technology." In other words, Not So Fast. Research is still trying to catch up with the effects of rapidly-changing technology—changes that have accelerated in response to the virus.

Extrapolating from Dunbar's recent book, *Human Evolution* (2016), I am concerned about whether, since the human brain is actually slightly smaller than previously measured and since, in Dunbar's words, "the social brain hypothesis provides us with a precise equation for predicting social group size from brain size" (p. 66), we humans can still be socially effective in a planetary group. And, probably much less effective on Facebook, on other social media sites, and in general, with our exploding on-line culture. That is, perhaps we can negotiate effective relationships in the real world, but not in a virtual reality mediated and infinitely extended by technology. Thus while we are evolving to be dependent on machines—in some cases, for our physical survival—we may be unequipped to handle the effects of information overload on social impulses such as altruism, parenting, and even sexuality. Do we feel the same urge to be generous towards someone who exists on a screen? Maybe, maybe not.

Randy L. Buckner, a neuroscientist and professor at Harvard, has wondered whether "the human brain is mismatched to the modern world in many ways." He worried about whether "brain systems specialized to tune social expectations through observation are now fed highlight reels of pretend lives on social media." That is, our brains are "tuned" to *real* people whom we can physically sense as real. The town meeting is the brain's ideal setting. In his Harvard course, "The Human Brain Then and Now," Buckner describes how different current societies are from those of stone-age hunter-gathers whose brains we largely inherited. Some of the "mismatches" Buckner perceives involve the consequences of "Social network sites [that] tap into innate social drives." The uses and abuses of current social networks have little to do with the social practices extant when our brains evolved.

The Center for Humane Technology features specific examples of what Randy Buckner is talking about. It states that "our society is being high-jacked by technology. Snapchat turns conversations into streaks, redefining how our children measure friendship. Instagram glorifies the picture-perfect life, eroding our self-worth. Facebook segregates us into echo chambers, fragmenting our communities." That is, natural human interactions have become simulacra of themselves. Thus, while everybody acknowledges the benefits of a wired world, we also have to step back and *discriminate* among even innocuous-seeming platforms can disrupt our natural social impulses. In part, I wrote this book to demonstrate just how profoundly our brain mechanisms are tied up with such impulses and, hence, how risky it could be to mess with them. To understand the neuroscience behind such impulses, built upon the eons of evolution that brought us to this place, is to realize that lightning-fast technological interventions (rivaling the speed of our own brains) could be massively disconcerting now and into the future.

I would also like to point out that these technological interventions are occurring at a time when some societies are already fragmenting. While some deviations from prosocial behaviors are due to war, there are many other sources of violence (Kleinfeld, 2018, p. 5). In 2016, homicides, organized criminal violence, killings by so-called "security services," and terrorism outdistanced war in accounting for violent deaths. Some states, according to Kleinfeld, have "violence as a governing strategy" (pp. 14–15). She suggests that "Privilege Violence" occurs in highly unequal, polarized societies governed by elites who twist the law to their benefit and monopolize state resources. These societies suffer "decivilization" (p. 73). Overall, she suggests, social trust is low—even in the United States. No one would deny that since 2016, our institutions have come under extraordinary stress from a norm-shattering federal government. My point is to ask why, in the midst of all this stress and disorientation, we should tolerate technologies that make us think we are connected when, in fact, our brains are probably ill-suited to the type of connection that is actually produced. The answer may be that, in the midst of the coronavirus, any connection is a good connection, and that is probably true. But were we becoming and have we now become inured to a virtual social life? That is the conundrum. Perhaps our immediate survival (dependent on technology) may be adverse to our long-term survival as optimally functioning social beings. At some point, we will need to think deeply about this question.

David Barash, the University of Washington biologist whom I cite in Chapter 2, also suggests that circumstances do not favor our successful management of social realities. His term, "anthrodiminution," refers to "the often painful but objectively unavoidable recognition that we're not as special as many people like to think" (Barash, 2018, p. 179). He distinguishes between *biological evolution* "a relatively slow-moving organic process… that bequeathed us a notably large and creative brain, with a mind to match" from *cultural evolution* which features the possibility of cultural "mutations," unpredicted changes that "can arise, be adopted, modified and passed along within days, without waiting for the passing of even a single biological generation" (pp. 180–181). Can we manage the social developments of Barash's cultural "mutations" with our current brains?

In a similar fashion, economist Niall Ferguson (2018) worries about the impacts of social media on social behavior. In our networked world, the situation is of "syndicating data outwards, not controlling what happens when it gets to the other end of the connection" (p. 351). In effect, data has a mind of its own, and is liable to produce *unintended effects*. Facebook, like various other social networks, was "not originally created to be a company. It was built to accomplish a social mission, to make the world more open and connected," (p. 354) to create "a service that was free to the user and unconstrained by geography or language." But now these users are approaching 2 billion (p. 355) with a worrisome quotient of misinformation and hate speech. Hence, "it was naïve to assume that we were witnessing the dawn of a new era of free and equal netizens," all initiating and responding to social communications in positive and truthful ways.

Referring to Renee DiResta, Ferguson cites our current "digital crowds," always seeking to grow and to do so without physical limitations; to have a degree of equality "but with higher levels of deception, suspicion and manipulation"; to achieve a high density because "digital identities can be more closely packed"; and needing direction which is offered by cheap clickbait. (p. 381). In Ferguson's telling, social networks such as Facebook have the danger of becoming *massively unstable.* How do we negotiate such networks without becoming overwhelmed? The so-called addiction that these networks produce is likely to reflect, at least in part, a futile effort to "keep up"—to not be left out and not miss anything. By way of contrast, in the real world, we move at the speed of clearly defined groups in a defined space that is more or less stable for a reasonable amount of time. See the difference?

In this regard, Ferguson notes that "homophily in social networks seems to result in polarization" (p. 382). Social problems can easily multiply and get out of control.

Part of the social instability feared by Ferguson and others results from the fact that social media comprise scale-free networks. This means that the distribution of social connections in such media resembles what applied mathematician Albert-Laszlo Barabasi describes with exponential equations, such that small perturbations at certain places in the network can explode into widespread and massive results throughout the network. Untrue claims can spread rapidly and widely (p. 387).

Obviously, I am inclined to share Dunbar's and Buckner's and Ferguson's views that our social brains may not be equipped to handle efficiently and usefully the "Library of Babel that is the Internet" (p. 389). "Never before have so many people been connected together in an instantly responsive network through which 'memes' (not Richard Dawkins' genes, but his cultural ideas) can spread even more rapidly than natural viruses" (p. 406). In Ferguson's view "The deepest social networks remain local and sociable" (p. 389). Ferguson's reference to viruses is deeply ironic, since it is the coronavirus in partnership with social media that has driven our increasing attachment to social media, and will do probably do so into the near future.

These authors are worried about the wholesale use of technology as a *replacement* for social contact on a human scale. They are astute observers of an evolutionary arc that seems seem detached (in its speed and general direction) from the known capacities of human beings to process information. "Forewarned is forearmed." If it is true that today's pervasive social media are stretching our social brain's abilities to perform as well as we expect, shouldn't we at least recognize the situation while it is still unfolding?

Today, neuroscientists are learning fast how the brain regulates behavior but, of course, social brain mechanisms are more multiple and complex than can be presented by a single book. The chapters above present facts and findings that we know for sure, but still underrepresent the brain's complexity. I reported on these facts because they fascinate me, but not just for that reason. By reporting some aspects of how the social brain functions, it is possible to try and estimate human adaptability into the future, assuming continuing stress. If we do not appreciate (and have respect for) how the social brain functions, we can abuse it like any other part of the body. If the book encourages you to think on the basis of well-understood data about our common neurobiology, then that's progress.

References

Barash, D.P., 2018. Through a Glass Brightly. Oxford University Press, Oxford.
Dunbar, R.I.M., 2016. Human Evolution: Our Brains and Behavior. Oxford University Press, Oxford.
Ferguson, N., 2018. The Square and the Tower. Penguin, New York.
Kleinfeld, R., 2018. A Savage Order. Penguin Random House, New York.
Pinker, S., 2011. The Better Angels of Our Nature. Viking, New York.
Rawls, J., 1993/2005. Political Liberalism. Columbia University Press, New York.
Ridley, M., 2010. The Rational Optimist. HarperCollins, New York.
Wrangham, R., 2019. The Goodness Paradox: The Strange Relationship between Virtue and Violence Inhuman Evolution. Pantheon, New York.

Further reading

Diamandis, P.H., Kotler, S., 2015. Abundance: The Future Is Better Than You Think. Simon & Schuster, New York.
Moore, S., Simon, J.L., 2000. It's Getting Better All the Time: 100 Greatest Trends of the Last 100 Years. Cato Institute, Washington.

Index

Note: Page numbers followed by *f* indicate figures.

F

G

H

Kolibri 360 GmbH
Ferdinand-Jühlke-Straße 7
99095 Erfurt, Deutschland
produktsicherheit@kolibri360.de